British Welding Research Association Series

WELDING PROCESSES

WELDING PROCESSES

P. T. HOULDCROFT, B.SC. (ENG. MET.) F.I.M.

Deputy Director (Scientific) British Welding Research Association

CAMBRIDGE AT THE UNIVERSITY PRESS 1967

Published by the Syndics of the Cambridge University Press
Bentley House, 200 Euston Road, London, N.W.1
American Branch: 32 East 57th Street, New York, N.Y. 10022

© Cambridge University Press 1967

Library of Congress Catalogue Card Number: 67–14286

Printed in Great Britain
at the University Printing House, Cambridge
(Brooke Crutchley, University Printer)

CONTENTS

PREFACE

Since primitive man first fastened a flint to a shaft the technology of joining has been a vital factor in the progress of civilization. Welding is the most adaptable of all joining methods and there are few things we handle nowadays which have not at some stage before reaching us depended on a welded joint. With its wide use in many industries a knowledge of the processes for welding is essential, not only to welding engineers and metallurgists, but also to designers, fabricators and users of welded products.

This book explains the basic principles and chief characteristics of welding processes, indicating the relationships between process variables and the function of particular techniques. There are numerous sources of information on the practical use of specific processes, many schools exist giving instruction in the practice of welding, and equipment for welding, the latest details of which can be obtained from the manufacturers, is under constant development. No attempt has been made, therefore, to reproduce in detail technique data sheets or to catalogue equipment.

An elementary scientific knowledge but no prior experience of welding has been assumed. The book is directed particularly to students, but because whenever possible recent research has been summarized it should also be of value to those with experience. For the specialist reader each chapter is concluded with a selected bibliography.

Many friends in industry have kindly supplied the author with data and illustrations. These organizations include: Arcos, A.1 Welders Ltd., BICC—Burndy Ltd., British Oxygen Company Ltd., British Welding Journal, Crompton Parkinson (Stud Welding) Ltd., E. I. DuPont de Nemours and Co. (Inc.), Engineering Ltd., General Electric Company Ltd., Iliffe Production Publications Ltd., Murex Welding Processes Ltd., Philips Welding Ltd., Rowen-Arc Ltd., Thermatool Corporation, Thermit Welding (Great Britain) Ltd.

The author is also indebted to his wife who typed and prepared the manuscript with great patience, to L. R. Parkes who read the manuscript and to his colleagues for helpful discussions. Finally, the book could not have been written without the British Welding Research Association. Not only did B.W.R.A. allow the author to use numerous illustrations from its published research reports but they also provided him with his experience of what is one of the most fascinating of modern technologies.

P. T. H.

1

PROPERTIES AND CLASSIFICATION OF WELDING PROCESSES

BASIC REQUIREMENTS

The ideal weld is one in which there is complete continuity between the parts joined and every part of the joint is indistinguishable from the metal in which the joint is made. Although this ideal is never achieved in practice, welds which give satisfactory service can be made in many ways. Not every welding process is equally suitable for each metal, type of joint or application, and much of the skill of the welding engineer consists in the recognition of the essential requirements which a particular weld must satisfy and the choice of the appropriate welding process.

Each welding process must fulfil a number of conditions. Most important, energy in some form, usually heat, must be supplied to the joint so that the parts can be united by being fused together. The heat may be generated by a flame, an arc, the resistance to an electric current, radiant energy or by mechanical means. In a limited number of processes such as pressure welding the union of the parts is accomplished without melting, but energy is expended in forcing together the parts to be joined and heat may be used to bring the weld region to a plastic condition. Fusion is generally considered as synonimous with melting, but in the context of welding it is desirable to distinguish at once between these words. By common usage the word fusion implies melting with subsequent union, and it is possible for the parts of a joint to be melted but not fused together.

Two surfaces can only be unified satisfactorily if they are free from oxide or other contaminants. Cleaning the surfaces before welding, though helpful, is not usually adequate and it is a feature of every welding process that the contaminated surface film is dissolved or dispersed. This may be done by the chemical action of a flux or the sputtering of an electric arc or even by mechanical means such as rupturing and rubbing. The contaminants which must be removed from the surface are of three types—organic films, adsorbed gases and chemical compounds of the base metal, generally oxides. Heat effectively removes thin organic films and adsorbed gases so that with the majority of welding processes where heat is employed it is the remaining oxide film which is of greatest importance.

Once removed, surface films and particularly nitrides, must be prevented from forming during the process of welding. In almost every welding process, therefore, there must be some way of excluding the atmosphere while the process is carried out. If a flux is used for cleaning the fusion faces of the joint, this also performs the function of shielding. If a flux is not used shielding can be provided by a blanket of an inert gas, or a gas which does not form refractory compounds with the base metal. The atmosphere may also be excluded mechanically by welding with the faces to be joined in close contact and the ultimate in protection from the atmosphere is obtained by removing it entirely by welding in a vacuum. Where the welding operation is carried out at high speed and with such limited heating that there is no time for appreciable oxidation, shielding may be unnecessary. It is possible with a few processes, however, for any contaminated molten metal to be expelled before the joint is completed or for the properties of the weld metal to be corrected by making alloying additions to the weld pool.

One further important requirement is that the joint produced by the welding process should have satisfactory metallurgical properties. In methods which involve melting of some part of the joint it is often necessary to add deoxidants or alloying additions, just as is done in the foundry. Frequently the material to be welded must have a controlled composition. Some alloys—happily few—are unweldable by almost any process, but a great many are only suitable for welding if their composition is controlled within close limits. These considerations are the basis of welding metallurgy, a detailed discussion of which is beyond the scope of the present book.

To summarize: every welding process must fulfil four requirements:

(1) A supply of energy to create union by fusion or pressure.

(2) A mechanism for removing superficial contamination from the joint faces.

(3) Avoidance of atmospheric contamination or its effects.

(4) Control of weld metallurgy.

TYPES OF WELDING PROCESSES

The simplest welding process would be one in which the two parts to be joined have their surfaces prepared to contours matching with atomic precision. Such surfaces brought together, in vacuum, so as to enable electrons to be shared between atoms across the interface could result in an ideal weld. The preparation of surfaces with this degree of precision and cleanliness is not feasible at present, although it is approached in space technology when metals may be in contact in the ultra-high vacuum of outer space. Slight rubbing of surfaces under these conditions can induce welding by satisfying the first two conditions above at limited points of contact, the third being supplied already by the vacuum. While such conditions of cleanliness and vacuum might be visualized for special micro-welding applications, alternative solutions must be found for practical welding.

The problem of achieving atomic contact between the parts to be joined is solved in one of two ways. Pressure may be applied so that abutting surfaces are plastically deformed giving the required intimacy of contact at least at asperities as indicated in fig. 1 *a*. The deformation also helps to satisfy the cleaning requirement by rupturing films. With ductile metals the plastic deformation can be accomplished cold but less malleable metals may be first softened by heat. Alternatively, the surfaces to be joined may

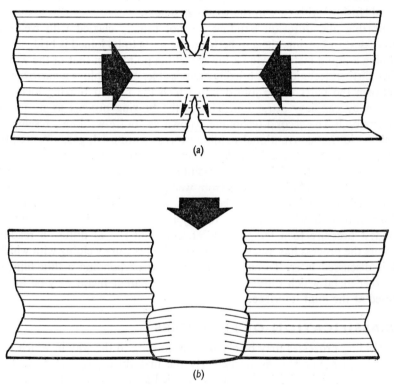

(a)

(b)

Fig. 1. Basic mechanisms of welding. (*a*) Union by flow.
(*b*) Union by molten metal bridging.

be bridged with liquid metal. The required adjustments in contour and structure are then effected as the melted metal solidifies (fig. 1 *b*). The majority of welding processes employ the latter method, and their variety is an indication of the many ways by which it is possible to generate locally the heat required for melting.

The two types of welding process described are fundamentally different, and the division between them forms the first breakdown in the classification of welding processes. Those welding methods employing pressure to plastically deform the faying surfaces are frequently called 'solid-phase' methods. There is no accepted term for the methods in which union is made through liquid metal but they may be called 'liquid-phase' methods.

For some years it has been customary to divide welding processes into 'pressure' and 'fusion' welding methods. See, for example, B.S. 499. The pressure-welding processes were those in which pressure is used at some stage in welding, and while this classification included all the methods which could be truly classed as solid-phase methods it also included several methods in which fusion takes place. This is a second reason for using the word 'fusion' with care in the welding context.

Brazing and diffusion bonding are not usually classed with welding processes. In brazing, the gap between the parts to be joined is bridged by adding a liquid metal having a lower melting point than that of the work. Two distinct techniques of brazing are possible, one employs a narrow gap between the parts to be joined into which the molten brazing metal is drawn by capillary attraction. This is the brazing process as normally understood. The second technique, known as 'braze-welding', resembles welding in that conventional welding processes are used, gas, arc, etc., to supply heat but no use is made of capillary attraction. It is only in the use of a filler metal with a melting point appreciably lower than the parent metal that the technique differs from welding.

Diffusion bonding is a specialized solid-phase joining method. In its simplest form in which the parts to be joined are heated in vacuum under modest pressure to a temperature at which interfacial diffusion readily occurs the process has much in common with pressure welding with which it is discussed later. To enhance the process of bonding a thin film of another metal which diffuses readily into the parent metal may be interposed at the joint interface.

CLASSIFICATION OF PROCESSES

Welding processes may be classified according to the way in which the four basic requirements—particularly the first three—are satisfied. The energy for welding is almost always supplied as heat so that divisions can be made according to the methods by which the heat is generated locally. These methods may be defined and grouped as follows:

(*a*) *Mechanical.* Heat generated by impact or friction or liberated by the elastic or plastic deformation of the metal.

(*b*) *Thermo-chemical.* Exothermic reactions, flames and arc plasmas. It is necessary to explain why plasmas should be put in the same class as oxy-fuel gas flames. Although chemical reactions may not take place in a plasma the method of heat transfer to the work is the same as for processes employing an envelope of burning gas. This holds for all processes in which the work does not form part of the arc circuit. The so-called non-transferred arc produces a plasma flame, whereas the transferred arc is a constricted arc and falls in the arc process category.

(*c*) *Electric resistance*. Heat generated by either the passage of a current introduced directly to the metal to be joined or by a current induced within the parent metal.

(*d*) *Electric arc*. Both a.c. and d.c. arcs with electrodes which melt and those which do not.

(*e*) *Radiation*. This category is suggested to cover the new processes such as laser and electron-beam welding and others which may yet be developed. The essential feature of a radiation process is that energy is focused on the workpiece and heat is generated only where the focused beam is intercepted.

It is not possible to define all welding processes completely by the source of thermal energy. This applies particularly to the many variations of arc welding and it is customary to complete the definition by reference to the way the process satisfies the condition of atmosphere control. All welding processes can be examined in the same way by placing the names of the processes within a grid formed by listing the sources of heat along one axis and methods of avoiding atmospheric contamination along the other axis as is done in fig. 2. The diagram can now be divided up into areas enclosing processes with a basic similarity. Seven such areas are readily identified corresponding to processes as follows: (1) solid-phase, (2) thermo-chemical, (3) electric resistance, (4) unshielded arc, (5) flux-shielded arc, (6) gas-shielded arc, (7) radiation.

Certain areas in the diagram can be marked out as regions where welding processes could not exist—for example flames cannot be used in vacuum.

This way of classifying welding processes is less rigid than the family tree method and makes it possible to account for certain anomalies. The resistance butt welding process, for example, while truly a solid-phase welding process, is normally included in the resistance welding category. In fig. 2 the position of this process is clarified by drawing the boundary of the group (1) solid-phase processes to include resistance butt and to exclude the remaining resistance processes. Similarly, electro-slag welding and its derivatives can be placed correctly in the resistance heat source grid, but may be linked with the flux-shielded arc processes with which they have a great deal in common.

There is no uniform method of naming welding processes. Many processes are named according to the heat source or shielding method, but certain specialized processes are named after the type of joint produced. Examples are stud, spot and butt welding. An overall classification cannot take account of this because the same type of joint may be produced by a variety of processes. Stud welding may be done by arc or projection welding and spot welding by electric resistance, arc, or electron beam processes. Butt welding may be done by resistance, flash or any of a number of other methods. Although in common usage many processes have abbre-

Welding process classification							
Source of heat		**Shielding method**					
		Vacuum	Inert gas	Gas	Flux	No shielding	Mechanical exclusion
No heat or heat by conduction		Cold pressure	Thermo compression bonding				Hot pressure Cold pressure
Mechanical		Explosive		1		Explosive	Friction Ultrasonic
Thermo chemical	Flames, plasma		Plasma	Atomic hydrogen	Gas	Forge	Pressure butt
	Exothermic reactions			2	Thermit		
Electric resistance	Induction			3		H.f. induction	Induction butt
	Direct				Electro-slag	Flash butt H.f. resistance Projection	Spot seam / Resistance butt
Electric arc	Consumable electrode		Inert gas metal arc	CO$_2$ metal arc Gas/flux metal arc	Covered electrode Submerged arc	Bare wire Stud Spark-discharge Percussion	
	Non-consumable		Inert gas tungsten arc	6	5	4 Carbon arc	
Radiation	Electro magnetic			7		Laser	
	Particle	Electron beam					

Fig. 2. Grouping of welding processes according to heat source and shielding method.

viated names, the full names often follow the pattern: first, a statement of the type of shielding (where mentioned); secondly, the type of heat or energy source; thirdly, the type of joint (where this is of specific and not general importance), e.g.

Inert gas	Tungsten-arc	Spot
(Unshielded)	Arc	Stud
—	Resistance	Butt
—	(Resistance)	Projection
(Vacuum)	Electron beam	—
(Flux-covered		
electrode)	Metal-arc	—
—	Friction	(Butt)

(Brackets enclose terms implied but not mentioned.)

It is often necessary when referring to processes to mention the way they are used, particularly whether the operation is manual or automatic. The practical operation of welding can be divided into three main parts:

(*a*) The control of welding conditions, particularly arc length and electrode or filler wire feed rate and time.

(*b*) The movement and guiding of the electrode, torch or welding head along the weld line.

(*c*) The transfer or presentation of parts for welding.

Processes are described as manual, semi-automatic, or automatic, depending on the extent to which the parts mentioned above are performed manually. Manual welding is understood to be that in which the welding variables are continuously controlled by the operator and the means for welding are held in the operator's hand. Semi-automatic welding is that in which there is automatic control of welding conditions such as arc length, rate of filler wire addition and weld time, but the movement and guiding of the electrode, torch or welding head is done by hand. With automatic welding at least parts (*a*) and (*b*) of the operation must be done by the machine. As feed-back control devices are introduced and welding takes its place more frequently in the automatic production line other definitions will be required.

THERMAL EFFECTS

With the processes in which heat is used the pattern of energy conversion to heat and its subsequent dissipation after welding is a major factor influencing the utilization of the process and the properties of the joint. The efficiency of energy conversion varies considerably between different welding processes with high overall efficiency being achieved with processes such as electric arc welding and probably the lowest with laser welding at approximately 1 per cent efficiency. Overall efficiency of energy conversion, however, is not usually of prime importance in the welding. Greater importance is attached to efficiency of heat transfer, energy level and energy intensity. These are the factors which influence welding speed and the extent of the heat affected zone in the workpiece. The former has economic implications, the latter, when excessive, causes distortion and often poor joint properties.

With gas welding heat is generated outside the workpiece and welding involves the transfer of heat across a boundary followed by conduction throughout the joint. This, with a minimum transfer efficiency of 30 per cent is clearly an inefficient process and very different from high-frequency resistance welding, for example, in which the heat is generated where required. The latter process is noted for its high welding speeds and narrow welds. Laser welding is inefficient in the conversion of primary energy (mains electricity) to the form in which it is used in the process (coherent light) and the utilization of the coherent light can be inefficient because of reflection. Such high-energy densities can be obtained, however, that in less than 10 ms weld pools can be established and boiled with a minimum spread of heat into the work. The low overall efficiency of laser welding takes its toll in the limited size of weld which can be produced with each pulse and the low frequency with which pulses can be made.

Not all the heat reaching the work is available for melting. Some heat must be used to build up the temperature gradient and there is a limiting heat input, depending on the thickness and thermal diffusivity below which melting will never occur because heat input cannot overtake heat losses. With high heat-input conditions, allowing high welding speeds, the heat dissipated in the workpiece is minimized but can never be reduced below half the total heat available. Inefficient processes or processes used at relatively low speeds result in losses three or four times greater than this. A measure of the efficiency of utilization of heat is given for the two-dimensional heat flow case by the weld characteristic devised by Wells. For a fusion weld this is a non-dimensional term $Vd/4\alpha$, where V is the weld velocity, d is the melted width and α the thermal diffusivity. The corresponding characteristic for a resistance spot weld is $d^2/4\alpha t$, where d is the spot diameter and t is weld time. In welds in which there is a high heat-input rate and the heat is therefore used efficiently $Vd/4\alpha$ will exceed 1. A low heat-input rate, allowing wasteful spread of heat, will be indicated by a weld characteristic of less than 0·1. For the majority of situations the characteristic lies between 0·25 and 1. Heat input is frequently expressed in terms of joules/inch of weld for a moving source or joules/second for a stationary source.

The relationship between the heat flow and weld parameters is given by the Wells simplification of the Rosenthal equation

$$Q = 8kT_m \left(\frac{Vd}{4\alpha} + \frac{1}{5}\right), \tag{1}$$

where Q is the heat input in cal/s/cm of plate thickness, k is thermal conductivity, T_m the melting temperature of the plate above the surroundings, α the thermal diffusivity, and $\alpha = k/\rho C$, where ρ is density and C is specific heat.

Using the above simplification Roberts and Wells (1954) subsequently derived a relationship giving the time t to cool between T_m and T_e for points on the line of the weld, where T_e is the equalization temperature or the temperature above the surroundings which a plate of finite size will eventually reach following welding:

$$t = \frac{d^2}{\pi\alpha} \left[\left(\frac{T_m}{T_e}\right)^2 - 1 \right] \frac{5(Vd/4\alpha)+2}{5(Vd/4\alpha)}. \tag{2}$$

From the above work expressions have been derived by Wells (1961) giving the time, t, to cool between two temperatures T_0 and T_1 on the line of the weld for the two-dimensional case of complete penetration in a single pass and the three-dimensional case of a bead on a plate:

$$\underset{\text{(two-dimensional)}}{t} = \frac{d}{V} \frac{5(Vd/4\alpha)+2}{4} \left[\left(\frac{T_m}{T_1}\right)^2 - \left(\frac{T_m}{T_0}\right)^2 \right], \tag{3}$$

$$\underset{\text{(three-dimensional)}}{t} = \frac{d}{V} \frac{5(Vd/4\alpha)+2}{8} \left[\left(\frac{T_m}{T_1}\right) - \left(\frac{T_m}{T_0}\right) \right]. \tag{4}$$

The chief difficulties in applying heat flow equations widely are that the change with temperature of physical constants is not always known and several metals have high latent heats of fusion. Energy absorption within the pool by latent heat and its subsequent release at the tail end of the pool on solidification is one reason why actual isotherms around a moving pool are more elongated than indicated by calculation. The Wells formulae give good correlation with experimental data for iron and can be adjusted to apply more accurately to metals having high latent heats by using a specific heat value corrected as follows:

$$\underset{\text{(corrected)}}{C} = C\left(1 + \frac{L}{CT_m}\right) \text{ cal/g/°C}. \tag{5}$$

Heat sink effects

The flow of heat into the workpiece is influenced not only by the heat input to the weld but also by the geometry and the material of which the workpiece is made. The capacity of a workpiece to absorb the heat available for welding is usually described as its heat sink.

A method of describing the heat sink of a joint has been devised by B.W.R.A. for use in calculating the cooling rate of welds in alloy steels. The total thickness of plate through which heat can flow away from the weld can be expressed in terms of the 'thermal severity number' (T.S.N.) of the joint. The unit of thermal severity has been taken as the heat flow through one thickness of $\frac{1}{4}$ in. plate, so that a joint with a thermal severity of 2 (T.S.N. 2) is obtained by depositing a weld in a position where heat can flow away from it through two thicknesses of $\frac{1}{4}$ in. plate (e.g. a $\frac{1}{4}$ in. butt weld). Two thicknesses of $\frac{1}{2}$ in. plate adjacent to a weld give a joint of T.S.N. 4, a fillet weld in a T joint in $\frac{1}{2}$ in. plate has T.S.N. 6 and so on. The T.S.N. is simply four times the total thickness of plate in inches through which heat can flow away from the weld, as illustrated by the examples given in fig. 3.

Type of joint and heat flow	Thicknesses of paths available for heat flow (in.)	T.S.N.
Through 2 plate thicknesses 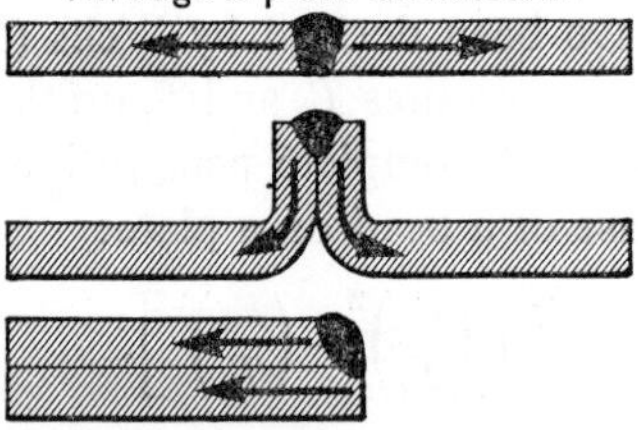	Both $\frac{1}{4}$	2
	$\frac{1}{4}+\frac{1}{2}$	3
	$\frac{1}{4}+\frac{3}{4}$	4
	Both $\frac{1}{2}$	4
	$\frac{1}{2}+1$	6
	Both 1	8
	$1+2$	12
Through 3 plate thicknesses 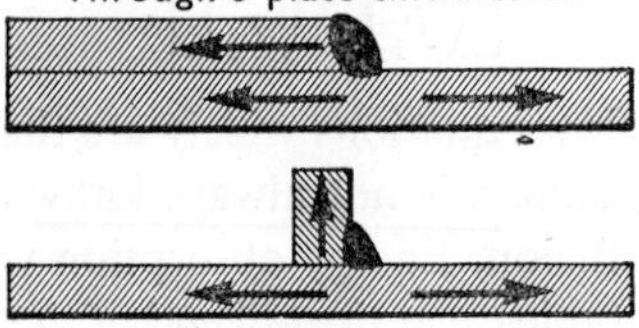	All $\frac{1}{4}$	3
	All $\frac{1}{2}$	6
	All 1	12
	All 2	24
	$\frac{1}{4}+\frac{1}{2}+\frac{1}{2}$	5
	$\frac{1}{2}+1+1$	10
	$2+1+1$	16
Through 4 plate thicknesses 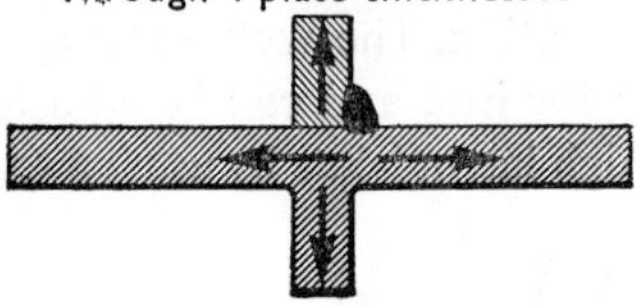	All $\frac{1}{4}$	4
	All $\frac{1}{2}$	8
	All 1	16
	All 2	32
	$\frac{1}{4}+\frac{1}{2}+\frac{1}{2}+\frac{1}{2}$	7
	$\frac{1}{2}+\frac{1}{2}+1+1$	12

Fig. 3. Determination of thermal severity numbers.

The above method cannot be applied to complex shapes or made to allow for the effects of jigging. An alternative method of assessment based on the Roberts and Wells calculation of the effect of bounding planes appears feasible. This work showed that in a plate with a half width of $10d$ (d being the melted width) the thermal cycles for a single pass with a weld characteristic between 0·25 and 1 are substantially the same as for a plate of infinite width. If it is required that only the upper part of the cooling curve should correspond with that in an infinite plate it is possible to reduce the plate width below $10d$. These observations lead to the idea that in considering the heat sink effect of a joint all metal lying farther away from the heat source than Sd can be neglected, where S is a factor chosen according to the part of the thermal cycle in which the interest lies.

Because metal near the heat source has a greater effect than the same mass a distance away, the heat sink effect must be related to the product of mass $x(Sd-l)$, where l is the distance of the mass from the source. Considering a cross-section of the joint the heat sink effect would be indicated by the sum of the product of individual areas and their 'significant' distance from the source

$$H_S = \Sigma A(Sd-l),\qquad(6)$$

S can be expressed in terms of the temperature range of interest giving

$$H_S = \Sigma A[3\cdot2(T_m/T)d-l].\qquad(7)$$

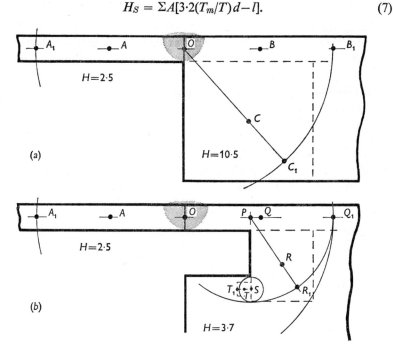

Fig. 4. Application of heat sink index. (*a*) Upper diagram. (*b*) Lower diagram.

The value of S has been chosen so that when expression (7) is applied to the extreme cases of two-dimensional heat flow in a single pass butt weld and the three-dimensional case of a bead on a plate of infinite thickness, the ratio of the heat sink indices approximates to the ratio of the cooling rates over the range T_m to T as determined from equations (3) and (4). Because temperatures are measured above the starting temperature the relationship (7) takes account of preheat.

Figure 4 gives an example of how the heat sink index might be used. Suppose that a comparison is to be made between the joints (*a*) and (*b*) on the basis of their ease of welding, the groove being machined in joint (*b*), with the object of giving welding conditions similar to those for the butt weld in equal thicknesses. The temperature of interest here is the melting point isotherm, so that $T_m/T = 1$ hence $S = 3.2$. For the left-hand side of fig. 4a the heat sink rating for a 0.5 cm thick plate where $d = 1$ cm is

$$H_S = a \times A_1 A = 2.5,$$

where a is the area of which A is the centre. Similarly, for the right-hand side of fig. 4a

$$H_S = b \times B_1 B + c \times C_1 C = 2.56 + 7.96 = 10.52,$$

and for the right-hand side of fig. 4b

$$H_S = q \times Q_1 Q + r \times R_1 R + t \times T_1 T = 2.56 + 1.13 + 0.01 = 3.7,$$

the heat path for area r being taken as OPR and that for area t as $OPST$. The provision of a groove in the joint has led to a substantial reduction in the heat sink effect, a 50 per cent difference being easily accommodated by the welding technique.

The physical properties of the workpiece are well known to have a major effect upon the heat sink. Although thermal conductivity is frequently regarded as the important parameter a more accurate indication of the relative capacities for conducting and absorbing heat is given by the thermal diffusivity. Table 1 indicates the thermal diffusivity of a number of common engineering materials. With a high thermal diffusivity metal such as copper, heat spreads rapidly into the work with less severe temperature gradients than for titanium, for example, which has a low value.

Table 1. *Approximate thermal diffusivity values for several metals*

Metal	C.g.s. units
Silver	1.7
Copper	1.1
Aluminium	0.9
Nickel	0.14
Steel	0.11
Titanium	0.06
Nimonic alloy	0.035

In welds between pieces of unequal geometry or thermal diffusivity the weld bead develops preferentially in the direction of least heat flow. If the inequality is too great for the heat input, fusion may not take place on the high heat sink side because the heat input there is insufficient to fulfil the minimum requirements. Raising the overall heat input, however, may only result in excessive melting on the low heat sink side. The remedy for this situation is (*a*) preheat the high heat sink side, or (*b*) machine the joint away on this side to reduce the thickness of metal, or (*c*) apply clamps on the low heat sink side to increase the cooling there and then to raise the overall heat input to compensate for the increased total heat loss.

Joining rate

The total cross-sectional area of metal joined in unit time is suggested as a useful practical method of comparing processes and their performance. For welds in sheet and plate this is conveniently the welding speed in inches per minute multiplied by the metal thickness. With fillet welds the factor would be speed × leg length and for spot welds the nugget area at the interface divided by time. For multi-pass welds the speed taken is that based on arc time to complete the joint. Each process and type of joint appears to have a characteristic joining rate independent to a large degree of thickness. Table 2 gives typical joining rates for a number of processes, and it will be seen that the high heat input rate processes generally give high joining rates. The joining rate for a given process and joint design frequently tends to start well below average with thin material, to rise as the thickness is increased and optimum welding conditions are established and finally to drop as the thickness becomes too great for the process to handle. Calculation of the joining rate can often reveal whether or not a process is being used to maximum capacity and anomalous results in a series are readily detected. The units of square inches joined per minute are convenient to remember and to use. A complementary calculation of energy input in terms of joules/in.^{2} is also instructive as it enables the efficiency of different processes to be compared.

Table 2. *Typical joining rates*

	in.^{2}/min
Oxy-acetylene welding, steel	0·15–0·5
Manual metal-arc, up to 0·5 in. steel	1·6–2·5
Manual metal-arc, over 0·75 in. steel	0·6–0·9
Manual metal-arc, vertical welds, steel	0·3–1·3
Submerged arc, steel	4–8
Submerged arc, high current techniques	10–13
Inert-gas metal-arc, backed butts, aluminium	3·2–3·8
Inert-gas metal-arc, as above, vertical	2·5
Tungsten-arc, butt welds, aluminium	0·8–1·3
High-frequency resistance, aluminium	60–130
High-frequency resistance, copper	35–60
High-frequency resistance, steel	60–90
Resistance spot weld, steel	4–10
Electro-slag (per wire)	1·5–2
Electron beam	Up to 25

Bibliography

American Welding Society. *Welding Handbook*, 5 vols.
Apps, R. L. and Milner, D. R. (1955). *Br. Weld. J.* **2**, no. 10, 475–85. Discussion, *ibid.* **3** (1956), 190.

Cottrell, C. L. M. and Bradstreet, B. (1955). A method of calculating the effect of pre-heat on weldability. *Br. Weld. J.* **2**, no. 7, 305–9.

Roberts, D. K. and Wells, A. A. (1954). A mathematical examination of the effect of bounding planes on the temperature distribution due to welding. *Br. Weld. J.* **1**, no. 12, 553–60.

Welding terms and symbols. B.S. 499, Part 1, 1965. British Standards Institution.

Wells, A. A. (1961). Oxygen cutting. *Br. Weld. J.* **8**, no. 2, 86–92.

Winterton, K. (1962). A brief history of welding technology. *Weld. and Met. Fab.* **30**, no. 11, 438–42; no. 12, 488–93; (1963), **31**, no. 2, 71–6.

2

THE WELDING ARC

GENERAL CHARACTERISTICS

The electric arc is the heat source for a variety of the most important welding processes, possibly because it is an easily produced high intensity source. It is, however, much more than just a source of heat. If required it can be arranged to transfer molten metal from the electrode to the work. It is also possible to use an arc simultaneously to supply heat and remove surface films—an important advantage where welding is done in the absence of fluxes. Within the envelope of the arc and the weld pool a whole range of complex gas–slag–metal reactions and other metallurgical changes take place.

An arc is an electric discharge between two electrodes which takes place through ionized gas known as 'plasma'. The space between the electrodes can be divided into three regions: a central region in which there is a uniform potential gradient and two regions adjacent to the electrodes in which the cooling effect of the electrodes results in a rapid drop in potential. These two regions are the anode and cathode fall, according to the direction of current flow. The length of the central region or arc column is influenced by the arc length but, where an arc is shortened to the extreme the influence of the electrodes may dominate the entire arc gap. With a high current arc at atmospheric pressure extremely high temperatures, from 5000 to 50000 °K, can exist in the axis of the arc column. From this central core the temperatures drop rapidly to the outer layers of the arc plasma and in a given atmosphere both the temperature and diameter of the central core depend on the current passed by the arc. Because most of the current is carried in this central channel and is influenced by its temperature and diameter, the relationship between arc current and potential is not according to Ohm's law but takes the non-linear form shown in fig. 5. When the current is increased from a low value the channel is enlarged and its temperature raised so that the potential drops to the minimum. In this part of the curve the arc is said to have a negative characteristic. The curve then remains flat or slightly rising until high currents are reached.

Welding arcs are generally operated in this high current region so that there is a tendency to a marked radial temperature gradient which, combined with the influence of the magnetic field created by the flow of current through the arc itself, exerts a constricting effect on the arc column—the pinch effect. The welding arc is also distinguished by its geometry which is invariably of the point-to-plane type. The spread of current from the electrode (point) to workpiece (plane) results in an axial component of the pinch effect and as the force is always from the point to the plane regardless of the direction of current flow, the ionized gas in the column is set in motion. As the force is proportional to I^2 it is to be expected that the movement of hot ionized gas or plasma jet would become of increasing signifi-

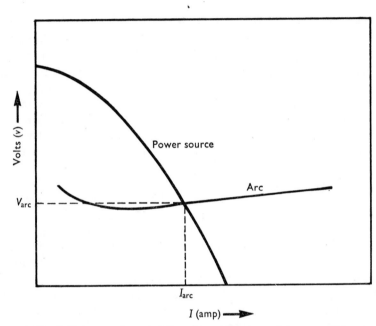

Fig. 5. Volt–amp characteristics of a welding arc and power source.

cance as the welding current is raised. The actual velocity of the jet is difficult to measure, but may be in the region of 10^4–10^5 cm/s. With a.c. the arc force builds up and decays with the current, but since it is independent of polarity the arc force pulses at twice mains frequency. Motion in the arc column implies a circulation of gas and it is known that entrainment of gas occurs in the region of the arc root on the electrode.

The behaviour just described is a generalization based on the arc root being larger in the plate than the electrode and the assumption by the arc of its well-known bell shape. In fact, the forces at work within the arc are greatly influenced by the behaviour of the arc roots and the type of emission taking place there. This is dependent on the composition of the electrodes and the gaseous atmosphere in which the arc operates.

The cathode of a welding arc has a major effect on the arc forces and the welding process. Materials such as carbon and hot tungsten emit electrons readily so that there is no well-marked arc root or cathode spot and the average current density over the emitting area is relatively low. This is thermionic emission. Where the arc root is at a well-defined and stationary cathode spot the cathode mechanism is described as plasma emission. This mechanism is associated with a contraction of the plasma at the arc root and relatively high current densities. A third type of cathode involves what is called field emission which takes the form of a relatively large cathode area within which many small cathode spots are active and in a constant state of movement. This activity is noticed particularly at the edge of the arc root where fresh supplies of metal oxides are available to encourage emission.

The essential feature to consider in relation to arc forces is the current path. If this is contracted through having to pass through a small arc root forces will be exerted, on both sides of the contraction, in liquid metal or arc plasma, away from the contraction. Similarly, any lack of symmetry within the system will result in deflection of the arc or liquid metal of the electrode due to the motor effect defined by Fleming's rule.

The practical importance of arc forces is in the way they detach and transport metal melted from an electrode to the workpiece. Drops are detached from electrodes by a mechanism which involves the electro-magnetic forces within the molten tip and the suction and drag effect of the plasma jet. Gravity is a dominant factor only at low currents when the forces referred to have low magnitude. The electro-magnetic forces do not always act in such a way as to detach drops. If current leaves the end of the electrode through the restricted outlet of a small anode spot the converg-ence of current within the drop results in forces tending to retain the drop on the wire.

TYPES OF WELDING ARC

From the welding point of view arcs are of two types according to whether or not the electrode is melted. If the electrode is refractory—that is if it is made of carbon or tungsten—it is not melted away in the process of arcing and is said to be non-consumable. When a lower melting-point metal is used the end of the electrode melts and molten droplets can be detached and transported across the arc gap to the work by the fast-moving plasma jet. The electrode when melted is said to be consumable, and because the detached droplets form part of the weld the material of the electrode is usually similar to that of the work. Any arc-welding system in which the electrode is melted off to become part of the weld is described as 'metal-arc'.

With a non-consumable electrode heat finds its way into the work by the electron or ion processes which take place at the boundary of the arc column with the work—this being the largest source of heat—and also by

2

the impingement of the hot plasma jet and the recombination of any gases dissociated in the column. Heat is lost to any fluxes present in the arc and also to a limited extent, usually only a few per cent, by radiation, and to the gases leaving the arc column. Additionally, heat generated at the column electrode boundary is lost by conduction up the electrode. If the electrode is melted and is transferred to the weld pool this heat is available again in the pool. Since the electrode is normally of limited diameter the passage of the arc current down the electrode to the arc root can cause resistance heating to an appreciable degree. In the consumable electrode or metal-arc processes this heat is also transferred to the pool. Because more of the heat developed by the arc is made available in the weld pool the con-

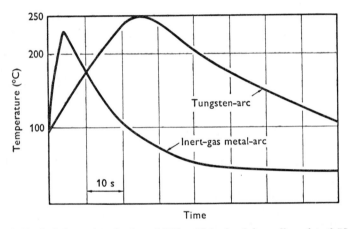

Fig. 6. Typical thermal cycles in a 0·25 in. thick aluminium alloy plate 0·75 in. from weld centre with non-consumable electrode (tungsten-arc) and consumable electrode processes.

sumable electrode arrangement has a higher thermal efficiency than the non-consumable electrode system giving narrower heat-affected zones as shown in fig. 6. Typical efficiencies are 75–90 per cent for the former and 50–60 per cent for the latter.

ARC INITIATION

An electric arc cannot be switched on merely by applying the potential required by the arc to the cold electrodes. The arc can only be ignited by first providing a conducting or ionized channel between the electrodes. This can be done in two ways at atmospheric pressure; by applying a sufficiently high voltage between the electrodes to cause a discharge or by touching the electrodes together and drawing them apart. Both solutions are adopted in practical welding according to the welding process in which the arc is being used.

Spark discharges are frequently used for igniting gas shielded arcs.

Voltages in excess of 10^4 V are required to break down the arc gaps used. Once breakdown has occurred, however, the voltage drops rapidly and the current begins to rise until after about 1 ms the normal arc voltage and the steady-state conditions of a stable arc are approached. Voltage and current continue to change slowly for seconds later as thermal equilibrium is achieved in both electrode and the work where the weld pool is formed. The use of voltages sufficient to break down arc gaps directly would be lethal in welding and it is therefore necessary to employ a high-frequency discharge. Spark gap oscillators are generally employed.

The most common method of striking an arc in welding is by touching the electrode on the work and then withdrawing the electrode as the contact area becomes heated. In welding terms this is known as a 'touch' start and an arc established by this method is said to be 'drawn'. The process of drawing an arc starts with the creation of an area of very high temperature at the point of contact which may cause local melting. As the electrode is drawn away from the work the molten bridge from electrode to the work is extended and is finally fused by the passage of the short-circuit current. Metal vapour may be generated and a transitory arc discharge takes place. If the power circuit is suitable a stable welding arc will then be established. The transitory arc discharge formed when the molten bridge is broken, which may not develop into a true arc, can occur even when the applied voltage is less than that required to run the stable arc. This is because there is inevitably considerable inductance in commercial power plant which leads to a voltage surge. The phenomenon is observed in the process of flash welding which is carried out at only a few volts.

In gas-shielded welding certain additions to the circuit are feasible to aid touch starting. With tungsten-arc welding the arc may be started at low power to avoid contamination of the tungsten and spatter on the work, while with a consumable electrode a current surge may be used to make the start more reliable when the process of withdrawing the electrode to draw an arc is not feasible. Small diameter electrodes such as those used in gas-shielded metal-arc welding generally fuse at the point of contact with the work so that the arc is initiated without the need to withdraw the electrode from the work.

ARC MAINTENANCE

Once an arc has been ignited and thermal equilibrium established it can often be re-ignited after a momentary extinction with relative ease. Thus, while several thousand volts might be required to ignite an arc with cold electrodes, only tens or hundreds of volts are required to re-ignite a thermionic arc. The presence in the electrodes of materials which are good thermal emitters is of considerable help. Re-ignition is a particular problem with the a.c. arc which is extinguished at the point of current zero on each reversal twice in every cycle. For the arc to re-ignite the required voltage must be available at this time of current zero. It is for this reason that in a.c. arc-welding plant the current wave is arranged to lag the voltage wave by using a low operating power factor of usually about 0·3. For the arc-welding transformer the power factor is given by the ratio of arc voltage to open circuit voltage

$$\phi, \text{ power factor } = \frac{\text{arc voltage}}{\text{open circuit voltage}}.$$

At the point of current zero under these conditions almost the full open circuit voltage of the transformer is available for re-igniting the arc as shown in the oscillogram (fig. 7). The operating power factor can be raised

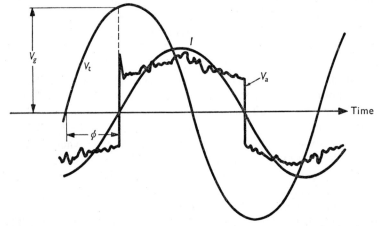

Fig. 7. Voltage and current wave-forms for an a.c. welding arc. V_t, no load voltage of transformer; V_a, arc voltage; I, arc current; V_g, voltage across arc gap on reversal.

while maintaining the ease of re-ignition only by using auxiliary means of re-ignition such as high-frequency, high-voltage oscillators or other devices to supply a pulse of high voltage at the appropriate instant. If such arc-maintaining devices are used the power factor can be improved and the open circuit voltage of the transformer reduced. These techniques are used with gas-shielded welding with a.c. since there are no fluxes or stabilizing agents present.

In d.c. welding it is only necessary to consider re-ignition after arc extinction due to accidental short circuiting or the process of metal transfer from the electrode. This requires a suitable dynamic characteristic from the power source so that voltage and current can recover their normal values rapidly giving what the welder knows as a smooth arc.

ELECTRODE POLARITY

Arc welding may be carried out using a.c. or d.c. with the electrode either positive or negative, the choice of current or polarity depending on the process or the type of electrode, the arc atmosphere and the metal being welded. With most common metals the metal transfer is more uniform, frequent and better directed when the electrode is positive. The positive pole of the arc becomes the hottest so that when using d.c. it would be preferable to employ the electrode negative (DCEN) polarity where the greatest fusion is required in the work. This is important with the non-consumable electrode process because the heat conducted up the electrode is completely lost to the welding process.

The arc has a most useful property of being able to disperse oxide or other refractory films on its negative pole. This arc-cleaning effect is particularly important in the gas-shielded welding of metals which form refractory oxides. Where arc cleaning is required the electrode must be positive (DCEP) if d.c. is used so that the negative pole on which the cleaning effect takes place is the work. In this way a clean weld pool is formed and a well-fused joint can be made. With a.c. the cleaning action takes place intermittently on the half cycles in which the work is negative.

Direct welding current requires the use of motor-generator or transformer–rectifier power plants which are more complicated than the transformer used for a.c. welding. In addition, with d.c. welding there is the phenomenon of arc blow. This occurs as a result of the flow of current in the work away from the weld pool which sets up magnetic fields that deflect the arc. A.c. welding would be preferred, therefore, were it not for the need of re-ignition on the change of polarity at each half cycle which introduces an arc-stability problem. This can often be overcome by the use of arc-stabilizing agents in the flux coverings so that in flux-shielded arc welding, especially of ferrous materials, there is usually a choice of a.c. or d.c. for welding. With gas-shielded welding a.c. is often chosen for tungsten-arc welding which employs a non-consumable electrode but not for gas metal-arc because smooth metal transfer from the electrode is achieved more readily with d.c.

ARC STABILITY

From the practical point of view great importance is attached to the property of arc stability. Unfortunately, there is no definition for this subjective quality which embraces a number of characteristics that are not all important in every circumstance. Experience suggests, however, several ways of identifying a stable welding arc.

At the electrode, the cathode or anode spot should remain essentially in the same position during the life of the arc and should not show short- or long-term variations in position such as jumping from point to point or moving slowly up the side of the electrode. If the electrode is consumable the transfer of metal should take place regularly, in small particles in an axial manner and without spatter.

On the workpiece the weld pool should move smoothly and remain in a fixed position in relation to the electrode. In fact, when an arc is struck between an electrode and a flat surface, the arc plasma should be normal to the surface and maintain this orientation when the arc is traversed. Sudden changes in direction of the plasma into new apparently 'stable' modes or the flickering of the arc root round the edges of a weld pool should not occur. This behaviour is sometimes evidence of non-uniformity of material or surface condition, often on a microscopic scale, or contamination of the gas shield. The arc seeks points of greatest emissivity and a uniformly oxidized surface can sometimes give better stability than one specially cleaned. Fast moving or small, low heat content weld pools are more susceptible to the effect of arc wander.

The arc-welding current should be steady, both in wave-form and value, and the arc should not be easily extinguished. Although the ease of striking of an arc from cold is strictly not connected with stability once hot, it is a desirable feature. Ease of maintaining the a.c. arc by high frequency or surge injection or self re-ignition is important. Finally the arc should not be sensitive to small changes in ambient conditions.

Bibliography

Bertung, M. (1964). Influence of d.c. generators and welding transformers on arc stability and electrode burn-off rate. *Br. Weld. J.* **11**, no. 4, 172–82.

Jackson, C. E. (1960). Science of arc welding. *Weld. J.* **39**, no. 4, 129–40; no. 5, 177–90; no. 6, 225–30.

Jennings, C. H. (1951). Dynamic characteristics of d.c. welding machines. *Weld. J.* **30**, no. 2, 117–38.

McCulloch, J. S. (1950–51). Electrification of shipyards (a.c. *vs*. d.c. welding). *Trans. N.E. Coast Inst.* **67**, 179–204.

Milner, D. R., Salter, G. R. and Wilkinson, J. B. (1960). Arc characteristics and their significance in welding. *Br. Weld. J.* **7**, no. 2, 73–88.

Needham, J. C., Cooksey, C. J. and Milner, D. R. (1960). Metal transfer in inert gas-shielded arc welding. *Br. Weld. J.* **7**, no. 2, 101–14.

Physics of the Welding Arc. Proceedings of Institute of Welding Conference, Princes Gate, London.

Somerville, J. M. *The Electric Arc*. London: Methuen.

3

FLUX-SHIELDED ARC WELDING

In this important series of welding processes the electric arc supplies the heat for fusion while a flux is responsible for the shielding and cleaning functions and often also for the metallurgical control. The most widely used form of flux-shielded arc welding is a manual process known as metal-arc welding.

METAL-ARC WELDING

Welding with arcs began in 1881 with the use by Auguste de Meritens of arcs from non-consumable carbon electrodes. Shortly afterwards in 1888 a Russian, N. G. Slavianoff, used a consumable bare steel rod as an electrode and he is generally accepted as the inventor of metal-arc welding. Manual bare wire welding, as it was known, was employed for almost half a century but is now virtually unused. Considerable skill was required to strike and maintain an arc and because the operation was done in air the deposited metal was seriously contaminated with oxygen and nitrogen. This affected the impact properties adversely.

Not long after the introduction of the bare-wire metal-arc process attempts were made to overcome these difficulties by coating the electrodes. These attempts may have been prompted by the fact that from the earliest days the importance of the surface of the electrode wire was recognized. Although described as 'bare' the wire used for welding frequently received a lime wash during wire drawing which resulted in the surface of the wire having a thin film of rust and lime. Although of negligible thickness this film led to a marked improvement in arcing properties. Kjellberg introduced the first flux-coated electrode in 1907 and clearly understood that the coating could have other functions than merely that of stabilizing the arc.

Since their introduction at the turn of the century metal-arc electrodes have been the subject of continuous development. In spite of competition from other more recently developed methods of welding the metal-arc process, in which electrodes in the form of short lengths of flux-covered rod are held by hand, has become the most widely used welding process. From the thin wash coating of some of the early electrodes which were applied by dipping wire into a slurry, the modern electrode has developed in which the flux is of considerable thickness and is applied by extrusion.

Electrode coverings

The covering on an electrode has several functions to perform—it must stabilize the arc, provide a gas and a flux layer to protect the arc and metal from atmospheric contamination, control weld-metal reactions and permit alloying elements to be added to the weld metal. Finally, the slag left behind on the surface of the weld should assist the formation of a weld bead of the proper shape. Once the weld is completed and has cooled the slag is no longer required and should be capable of being removed easily and quickly. As would be expected, there is now a variety of types of coating, each type being better suited for some applications than others. Because the characteristics of an electrode and the composition and properties of the deposited metal can be altered readily by the way the flux covering is compounded the preparation of electrodes for specialized purposes is possible. Electrodes are judged by the quality of the metal they deposit, the economy with which they deposit this metal and above all by the ease with which they can be used by the welder. The development of electrodes to meet the many demands made by the welder, welding engineer and inspection authorities is still largely a matter of skill and experience.

Of the many ingredients in the covering of electrodes for welding mild steel the most important are probably cellulose, generally as a chemically disintegrated form of wood pulp known as alpha flock; titanium oxide usually in the natural form of rutile; mineral silicates; iron oxides; basic carbonates such as limestone; fluorspar; ferrosilicon and sodium silicate. Iron powder is also incorporated in many modern electrodes. The behaviour of an electrode is determined, not only by the chemical constitution of the coating but also by source of supply, state of division and processing of the constituents. After the selected materials have been mixed and kneaded into a stiff paste with a binder, often sodium silicate, the paste is extruded round the low carbon rimming steel core wire which has been cut into suitable lengths and straightened. Subsequently the covered electrodes are dried and baked in continuous ovens to a controlled moisture content.

Where it is desired to deposit an alloyed weld metal ferro-alloys are added to the flux covering. With highly alloyed metal or non-ferrous metals, however, all or at least a major part of the alloying addition is contained in the electrode core wire.

Types of covering

In the United Kingdom electrodes for mild steel are divided into classes according to the type of flux covering they carry. A class 1 covering has a high cellulose content. This burns away to produce copious supplies of hydrogen and carbon monoxide which shield the arc from the atmosphere.

The presence in the arc of these gases with high ionization potentials results in a high arc voltage and, therefore, high arc energy which is responsible for the rapid burn-off and deep penetration of this type of electrode. During the deposition of cellulosic electrodes there is a tendency for decomposition of the organic coating constituents because the electrode core wire becomes heated by resistance. This results in a slight drop in arc voltage and an increase in the amounts of manganese and silicon finding their way into the weld metal. This effect is almost entirely controlled by the way the coating is compounded. As much of the coating is carbonaceous there is little slag left on the weld and this fact, together with the strong plasma jet giving a forceful arc, makes the electrode suitable for welding in all positions. It is widely used for pipe welding and structural work often with a vertical down welding technique where the weld metal is supported by the plasma jet. The absence of stabilizers in the coating and the high arc voltage necessitate the use of d.c. and an electrode-positive polarity.

Electrodes in classes 2 and 3 both contain titania (often as rutile) in the coverings which with a high content of ionizers makes the electrodes easy to use. A class 2 covering gives a viscous slag which supports the molten weld metal making the electrode suitable for horizontal–vertical fillet welds. Class 3 coverings contain added basic compounds and give a more fluid slag. They are therefore suitable for all positional welding giving medium penetration and a smooth welding arc. Because of the rutile and ionizers in the coating the electrodes may be used on a.c. or d.c. with either polarity.

Electrodes in class 4 are covered principally with oxides or carbonates of manganese and iron, with some silicates. This type of covering gives a fluid voluminous slag resulting in a clean smooth weld bead from which the slag is easily removed. The occurrence of trapped slag in deep multi-pass welds is reduced so that these electrodes are particularly suitable for high quality work where radiographic inspection is employed. As the slag is fluid only downhand welding is usually employed with these electrodes while the volume of the slag to be melted requires that the electrode positive polarity be used to take advantage of the extra heat at the electrode tip. A.c. can also be used.

The main constituent of class 5 coverings, now little used, is iron oxide which gives a heavy slag, low penetration and smooth concave fillet welds. The metal deposited is of lower strength than with other classes of electrodes but because of the smooth finish these electrodes were used where appearance was of first importance.

Probably the most important electrode type from the metallurgical point of view is class 6. These electrode coverings contain considerable proportions of calcium carbonate and fluoride as limestone and fluorspar. The electrode covering is produced with a very low moisture content so that the hydrogen content of the deposited metal is usually less than with other types of electrode and can often be as low as 5 ml/100 g. Table 3 gives typical hydrogen contents for weld metal deposited by electrodes of classes 2 and 6 as determined by the B.W.R.A. method. Electrodes of this class are called basic or low-hydrogen types though it should be noted that not all low-hydrogen electrodes are strictly of the basic or lime fluorspar type.

Table 3. *Hydrogen contents of mild steel weld metal*

Electrode and process	Hydrogen content (ml/100 g)		
	Diffusible	Residual	Total
Metal-arc			
a {Class 3, rutile, undried	25–35	1·5–5·0	27–37
a {Class 3, rutile, dried at 500 °C	18–22	1·5–5·0	20–24
b {Class 6, low hydrogen, undried	15–17	0·5–0·9	15·5–17·5
b {Class 6, low hydrogen, dried at 175 °C	11–13	0·4–0·6	12–13·5
c {Class 6, low hydrogen, undried	9·5–11·0	0·6–1·2	10–12
c {Class 6, low hydrogen, dried at 400 °C	3–4·5	0·5–0·8	4–5
Submerged-arc			
Damp flux	14–15	0·5	14·5–15·5
As received flux	8–8·5	0·3	8–9

Diffusible: hydrogen which escapes at room temperature. Residual: hydrogen remaining after storage at room temperature. Determinations made by B.W.R.A. methods—see F. R. Coe, *J. Iron and Steel Inst.* (Special report, no. 73, 1962).

Partly because of the low hydrogen content of the weld metal these electrodes are suitable for welding low alloy steels susceptible to heat-affected zone cracking. The weld metal has a high resistance to hot cracking and fissuring and is satisfactory for welding thicker steels and steels with higher carbon contents than other types of electrode. In addition, the weld metal has excellent mechanical properties—particularly impact properties. Basic electrodes are possibly not as easy to use as some of the other types, but nowadays they can generally be used in all welding positions on both a.c. and d.c. electrode positive. The slag is relatively fluid and not as voluminous as with the rutile type. Because they are used for high-quality applications for which they rely on the low moisture content of the covering these electrodes should be carefully stored and dried. For particularly severe applications such as the welding of alloy steels additional baking is used immediately before welding.

The introduction of iron powder into the covering of the electrode has a marked effect on its performance. Additions of iron powder range from 5 to 50 per cent in British electrodes, although the 5 per cent addition is not reckoned to put an electrode in the iron-powder class. Iron powder is added to the coating for two reasons, to give an increased deposition rate and to improve the behaviour of the arc. With conventional electrodes the welding current is carried wholly by the core wire, but with iron powder in the flux the covering becomes conducting near the arc thus providing an alternative path for the current. As a result the arc is spread out, tends to flare and deposits over a wider area with less penetration. The alternative current path in the arc area limits the current surge when metal globules short circuit between electrode wire and work so that spatter is reduced. These effects provide in practical welding terms a smoother, more stable arc, giving improved sidewall fusion, flatter welds and less undercut.

To take advantage of the higher deposition rate conferred on an electrode by iron powder it is necessary to make the maximum addition of 50 per

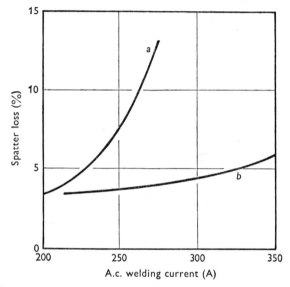

Fig. 8. Welding current *vs.* spatter curves for $\frac{3}{16}$ in. metal-arc electrodes. (*a*) Normal low hydrogen coating. (*b*) Similar coating with an addition of 50 per cent iron powder. (After Mathias.)

cent. Above this figure the behaviour of the electrode begins to deteriorate as the covering fuses away unevenly. The higher deposition rate of an iron-powder electrode depends not so much on the fact that extra metal is added from the covering as that it is possible to use higher welding currents for the same gauge of core wire. This is clear from an examination of the current/spatter curves in fig. 8. As the welding current is increased the amount of spatter also increases, but the curves for iron-powder electrodes rise less steeply than those for conventional electrodes so that higher operating currents are permissible.

The combination of rutile with large additions of iron powder in a heavy covering gives a particularly useful electrode type known as class 9. As it is manufactured to give a low hydrogen deposit this electrode combines many of the advantages of the lime-spar or basic electrode with the excellent arc behaviour of the rutile electrode and the enhanced electrode efficiency of the iron-powder electrode. A metal recovery of up to 150 per cent is claimed with the ability to weld in all positions, a feature not normally possible with the fluid rutile slags. The metal deposited by these electrodes is usually both sound and of good mechanical properties so that they may be used for heavy fabrications as well as for welding sheet where their ease of striking and lack of spatter are a considerable advantage.

Behaviour and use of electrodes

Electrode coverings are fused by the heat of the arc and the fused flux passes through the arc column as individual particles and as a covering to the transferring metal particles. If the covering has been uniformly compounded and extruded concentrically it will burn away uniformly round the periphery. For a number of reasons, including the thickness and constitution of the covering, the electrode core wire melts back behind the covering, which forms a cup or shroud around the end of the wire. With classes 4 and 9 electrodes this cup is well developed and allows the electrode to be used with what is known as a touch technique. The welder drags the electrode along in contact with the work and the depth of the cup is sufficient to maintain a suitable arc length.

The coverings of non-ferrous electrodes are of different composition from those of ferrous electrodes. Generally, there is no attempt to make additions to the weld metal through materials added to the covering as the core wire itself has the desired composition. Non-ferrous electrodes are used on d.c. electrode positive polarity for several reasons. This is the electrode polarity on which the arc itself is able to exert a cleaning action on the weld pool, which might be significant since many non-ferrous metals form refractory oxide films. It is also the polarity on which the greatest heat is developed in the electrode. The extra heat may be required to fuse the thick coverings of flux on high conductivity metals. Most important of all, however, with the electrode positive polarity the transfer of metal is smoothest with smaller better-directed particles than with the electrode negative polarity. This is probably because when the cathode is formed on the electrode the arc root tends to be localized so that the electro-magnetic forces do not assist drop detachment. With the electrode positive the arc anode will frequently envelop the end of the wire giving favourable conditions for transfer. These considerations are particularly relevant for gas-shielded welding.

Metal is transferred from a metal-arc electrode by one of two mechanisms or a combination of both. A spray of droplets of varying size may flow from the electrode in such a manner that their passage does not greatly disturb the arc, as shown in fig. 9*a*. Alternatively, drops may grow in size and short circuit the arc gap. Transfer of the drop then takes place by a process of bridging as fig. 9 *b* indicates. Which of the two mechanisms predominates depends on the type and size of the electrode and the arc length.

Oscillograms of arc voltage and current reveal certain characteristics of metal transfer and the electrode. Plate 1 *a* is a record of current and voltage for a cellulosic class 1 electrode used on d.c. The voltage record shows high-speed fluctuations because of the motion of the arc and the molten end of the electrode, which may be associated with the transfer of drops. A cyclic variation is observed in which the voltage rises after a short circuit to a value above average and then decreases until another short circuit or

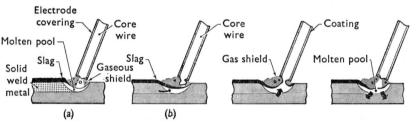

Fig. 9. Metal transfer in covered metal-arc electrode welding. (*a*) Spray or shower of droplets. (*b*) Stages in transfer by bridging.

series of short circuits takes place. Plate 1 *a* shows two such cycles, and these may be repeated between 2 and 6 times per second. The period of the cycles is influenced by type and size of electrode, the arc length and possibly also on operator characteristics. Voltage oscillograms for a.c. metal-arc welding show a similar pattern. Records for voltage and current for the same cellulosic electrode on a.c. are given in pl. 1 *b*. The voltage again fluctuates at high speed on each half cycle because of motion of the electrode end. After each reversal, the voltage rises rapidly to open circuit forming marked re-ignition peaks and this is the reason why cellulosic electrodes require high open circuit voltages and are often used on d.c. Occasional short circuiting when a drop is transferred by bridging is also observed.

In contrast, the record for a class 2, rutile-type electrode (pl. 1 *c*) shows an almost complete absence of re-ignition peaks, an indication of the smooth running of this electrode. Transfer by bridging is shown by the periods of near-zero voltage, and this type of metal transfer is particularly common with the rutile-covered electrode. Records for the class 6, lime-spar electrode are similar but there are more frequent instances of a sharp voltage increase towards the middle of the voltage half-cycle. This could be a result of a sudden excursion and lengthening of the arc or more likely the transfer of a large drop.

The transfer mechanisms described were examined by Sack amongst others as long ago as 1939, and fig. 10 shows two of his suggested processes for metal transfer. Bridging takes place smoothly under an envelope of slag as indicated in fig. 9*b*. Under the action of surface tension and the force of the arc formed when the bridge is broken the weld pool oscillates and assists the process of transfer. This mechanism can take place against gravity and occurs in vertical and overhead welding. The slag has a

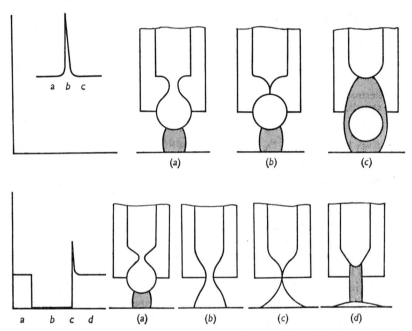

Fig. 10. Metal transfer-mechanisms as suggested by Sack in 1939. Upper globular transfer, lower bridging. Idealized voltage record at left.

dominating influence on the process of transfer from its effect on arc behaviour as well as its surface tension. Unfortunately, while the different classes of coating can be shown to have noticeably different surface tension properties as a function of temperature, it has not been possible to utilize this information on a quantitative basis.

Welding techniques for manual welding

Metal-arc welding electrodes are made with core-wire diameter from 18 s.w.g. (0·048 in.) up to ⅜ in. diameter. For all but exceptional circumstances, however, the useful range is 12–4 s.w.g. (0·102–0·24 in.). The length of each electrode depends on the diameter, for small diameter electrodes where manipulation of the electrode calls for the greatest control the electrode may be only 9–12 in. long. Generally, however, electrodes are manufactured to a length of 14–18 in. and are consumed at a burn-off rate of 8–10 in./min. At one end of the electrode the coating is removed during manufacture so that it may be gripped in the electrode holder through which the welding current is introduced.

The working current range for a number of electrode sizes is indicated in fig. 11. For economic reasons the welder should use the largest diameter of electrode suitable for each application. Because positional welding requires precise control of a small weld pool smaller sizes of electrode are

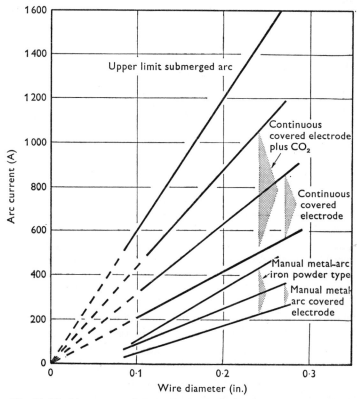

Fig. 11. Working ranges of electrodes for manual metal-arc and automatic processes of welding mild steel.

used for this purpose than for downhand welding where the pool is shaped by gravity. In multi-pass welds in fillets or grooved joints the first pass is usually laid using a smaller diameter electrode to obtain better access and root penetration. A typical sequence for a multi-pass downhand weld in thick plate is shown in fig. 12.

The welder's task is to direct the arc into the joint so that metal is deposited where required and to manipulate the electrode by whipping or weaving so that the arc force holds the metal in position and sweeps aside the slag. An electrode is never held perpendicular to the joint but is usually inclined so that an angle of about 110° is subtended between the weld bead and the electrode. This is sufficient to allow the welder to see the crater beneath the arc and for the arc force to prevent the undesirable flow of slag ahead of the advancing crater. As the process is intermittent a welder will usually have to stop and fit a new electrode many times in the course of making each weld. If the arc is broken by the electrode being withdrawn at the end of each electrode the weld crater will solidify without being fed by liquid metal and may form a crater pipe. To avoid this the arc is broken

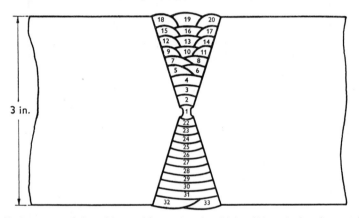

Fig. 12. Sequence of depositing weld runs in 3 in. thick mild steel plate by metal-arc. Typical of downhand welding: run 1 with 8 s.w.g. electrode; runs 2–14 with 6 s.w.g. electrode; runs 15–20 with 4 s.w.g. electrode. Typical of vertical welding: runs 22–27 with 8 s.w.g. electrode; runs 28–31 with 6 s.w.g. electrode; runs 32, 33 with 8 s.w.g. electrode.

by drawing the electrode slowly back along the bead while at the same time lengthening the arc. Before the next electrode is used the slag which has solidified over the crater must be chipped away to avoid slag inclusions. The arc from the new electrode is struck ahead of the crater and moved back to take up the end of the previous bead. The stop/start position at electrode changes is a serious source of slag inclusions, porosity and lack of fusion. This part of the welding technique must be mastered thoroughly if the welder is to produce quality work.

Current, voltage (arc length) and speed are important process variables. Low current will give an irregular weld bead which sits on top of the plate; a high current will give good fusion but with excessive spatter. Short, low-voltage arcs give irregular beads with poor penetration and a tendency to slag inclusions; high voltages and long arcs result in spatter and a tendency to pick up nitrogen from the air giving porosity. High welding speeds result in peaky, undercut beads; low welding speeds cause broad beads which tend to overlap on the work.

Applications

Manual metal-arc welding is the most used welding process. Because it can be used in all welding positions with mild, alloy, heat and corrosion resisting steels and with almost equal success with some copper and nickel base alloys, the process is widely used in shipbuilding, structural and general engineering. Almost ten times as many mild steel electrodes are used as all other special types put together. It is a low capital cost process which uses portable equipment and can be applied to a wide range of joint types. However, the cost of the metal deposited on mild steel is over thirty times that of the base metal per ton.

Joints such as those between attachments and main structures, joints in pipe and in complex assemblies of plate and section are difficult to mechanize and are ideal for manual metal-arc welding (pls. 2 and 3).

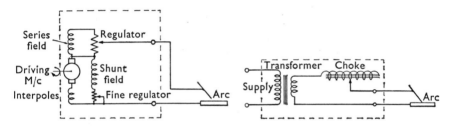

Fig. 13. Simplified circuit for metal-arc welding: (*a*) d.c. motor-generator; (*b*) a.c. transformer.

Power for the arc

The power source which supplies the current for metal-arc welding may be either a.c. or d.c. In the early days of welding d.c. was used because of difficulties in stabilizing the a.c. arc. With the adoption of the covered electrode both types of power source are used, although there are certain national preferences which have arisen for historical reasons, for example direct current is more generally used in the U.S.A. and certain Scandinavian countries than in Britain.

Regardless of the type of power source, means must be provided for controlling the current delivered to the welding arc. Figure 13 shows two arrangements for metal-arc welding. Because the process is manual an extremely precise control of arc gap and hence voltage is not possible. Voltage fluctuations must be accepted and the power source must be such that a swing in voltage should not be accompanied by a large corresponding change in current. This is important, particularly where a constant heat input is required to give control of the weld penetration as in welding thin sheet and for root runs.

The behaviour of the power source is given in part by its volt–amp

3   HWP

characteristic curve. Simplified curves are shown in fig. 14, curves 1–3. The intercept on the voltage axis is determined by the open-circuit voltage setting, that on the current axis by the short-circuit current which is limited by the resistance or impedance of the power circuit. During welding the current which is drawn by the arc depends on where the volt–amp curve for the generator is intersected by the similar curve for the arc. As explained in chapter 2 there is a family of curves for each arc type, each curve giving the conditions for a particular arc length (fig. 14, l_1–l_4).

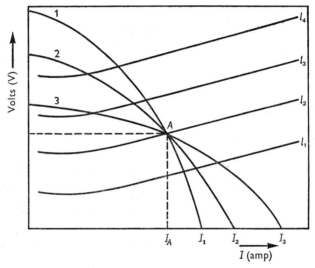

Fig. 14. Volt–amp characteristics for an arc of increasing length (l_1–l_4) and power sources of decreasing slope (1–3).

When the arc gap is increased by the welder the conditions at the arc are given by progressively higher arc characteristic curves which intersect the power source characteristic at increasingly higher levels. The voltage across the arc rises and the welding current drops. A decrease in arc gap reverses the process.

Power sources for manual metal-arc welding generally have volt–amp characteristic curves which fall with increasing current, and these are said to be drooping chracteristics. The shape of the curve depends on the design of the power source and in particular on the way in which the current is controlled. Because these curves describe the conditions when equilibrium has been attained they are often known as static characteristic curves and they give no information on the way the power source would behave under transient or dynamic conditions.

In a.c. welding plant the current is usually supplied from a single-phase transformer and is regulated by fixing the short-circuit current by adjusting the reactance of the secondary circuit. This can be achieved in a great variety of ways, for example a constant voltage transformer may be employed with a separate choke or reactor in the secondary circuit. The reactor may be tapped, or made adjustable magnetically, either by mechanically altering the iron in the core or by saturating the core with a variable direct current. This latter method is usually called a saturable reactor. Alternatively, a single iron core incorporating both transformer and reactor is used. The current is limited by mechanically altering the arrangement of the core, as in the leaky reactance type of transformer or by moving the coils about a fixed core.

Direct current can be supplied by a transformer–rectifier system in which a rectifier, usually of the selenium plate type is added to one of the a.c. devices just described. Another popular method of supplying d.c. power for welding is the motor generator set in which a d.c. generator is usually driven by a three-phase induction motor of mains frequency. Both separately and self-excited compound generators are used. The output characteristics and current control are varied by the excitation of the generator through adjustment of the shunt field and by control of the series field. It is generally possible to obtain a wider range of output characteristics with motor generator than with transformer–rectifier sets and this makes them particularly popular where a variety of applications of metal-arc welding must be attempted. The steeply drooping characteristic, for example, is favoured where only minimum variation in current can be tolerated, as in welding thin sheet. For positional welding, however, the operator must be able to vary the current by the way he controls the arc length so that a flatter characteristic is then preferred.

Because the welding current delivered by a power source depends on where the arc characteristic meets the static power source characteristic, and both these can be varied considerably, it is difficult to preset the power source to deliver precisely a particular current. Power sources have their controls calibrated for average mild steel electrodes and normal arc lengths, but these settings give only an approximate indication of the arc current. It is always advisable to check the current with an ammeter if unusual electrodes or techniques are being used.

Deposition rate. The efficiency of metal-arc electrodes is often judged in terms of the weight of metal deposited in unit time—the deposition rate (usually expressed in lb/h as this simplifies calculations of productivity); mg/s units are also used. As some of the electrode is lost as fume or spatter the electrode melting rate (mass per unit time) never equals the deposition rate. This discrepancy is described by the electrode efficiency which is the ratio of mass of metal deposited to the mass of core wire melted. With an iron powder electrode metal is also added from the covering so that, in spite of losses, electrode efficiencies considerably in excess of 100 per cent are quoted. The linear rate of consumption of an electrode is the burn-off rate, expressed in in./min or cm/s. Burn-off rates depend mainly on current but are influenced by electrode polarity, presence of flux or gas, surface treatment of wire, and the amount of resistance heating in the wire.

Mechanized welding with covered electrodes

Two of a number of methods which have been tried in an attempt to dispense with human agency in depositing covered electrodes are worthy of mention.

In 'firecracker' welding a long electrode of the class 4 or touch type is placed in the joint and held in position by a grooved copper bar. An arc is ignited by short circuiting the work to the exposed end of the electrode. The arc length is controlled by the thickness of the electrode covering and the depth of the cup formed in the electrode by the arc. Although electrodes several feet long can be used, 'firecracker' welding is not truly continuous and is limited to single-run deposits. Because the welding speed is lower than can be achieved by automatic methods of welding the process does not find much application and is limited to fillet welding in positions inaccessible to conventional automatic or manual methods.

A method of continuous welding with metal-arc electrodes was provided by an automatic rod-feeding head. This equipment had a magazine from which electrodes were fed one at a time to a vertical endless belt so that they could be burnt away in succession. The practical advantage of this equipment was that special electrodes already available for manual use could be deposited automatically without the necessity to develop the special coverings normally required for other automatic processes. Although the time taken by the manual welder in deslagging after depositing each electrode and in fitting a fresh electrode was eliminated, welding currents were not significantly higher and the welding speeds could not compete with other methods of automatic welding.

AUTOMATIC FLUX-SHIELDED WELDING

In metal-arc welding with covered electrodes the current must pass through the core wire, and there is a limit to the length of electrode which can be used because of the resistance heating effect of the current in the core wire. The process is essentially intermittent and its adaptation to a continuous form is the main difficulty to be overcome in using the process automatically. Several successful solutions have been developed to the problem of automatic continuous welding with a flux-shielded metal-arc. Each overcomes the difficulty of introducing the current to the electrode wire at a point near the arc in a different way. This is an essential feature of an automatic process and enables high welding currents to be employed. Three methods are illustrated diagrammatically in fig. 15.

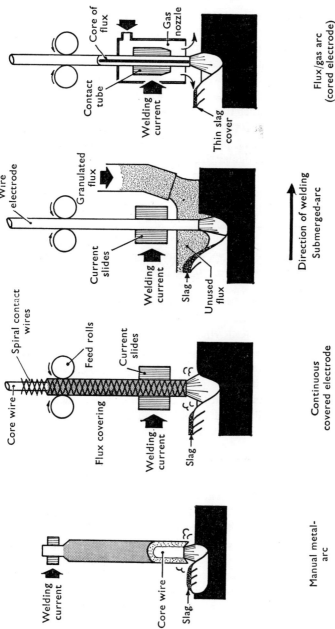

Fig. 15. Manual and automatic flux-shielded welding.

Continuous covered electrode

If the electrode wire is continuously covered with a flux some way must be found of making electrical contact with the electrode wire through the flux. A way of doing this is shown in fig. 16, in which the core wire is helically wrapped with both left- and right-handed spirals of wire. The coating fills the spaces between the spirals. Current contact slides in the welding head touch the parts of the wire spiral which protrude through the coating and conduct the welding current through to the core wire. Welding currents from 200 to about 1 000 A can be used depending on the gauge of electrode (fig. 11, p. 31). As with metal-arc electrodes the current range over which a particular size of electrode is usable depends to a large extent on the way the flux covering is compounded. With the continuous

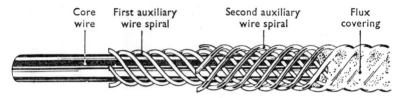

Fig. 16. Construction of a continuous covered electrode.

covered electrode there is also the limit set by the inability of the wire spiral to carry more current without risk of overheating and collapse. Careful design of the electrode and wire spiral and the use of long current slides overcomes this difficulty.

It is necessary to ensure that the flux covering is of sufficient mechanical strength to survive coiling and uncoiling and handling. There is also a limit to the thickness of the flux because of increasing the problem of introducing current to the wire.

By the addition of a shield of carbon dioxide the protection of the weld can be greatly improved, there is greater freedom in compounding the flux and the permissible current level is raised. Along with these improvements goes greater penetration and welding speed. Both these variations of the process belong to a class known frequently as 'Open-arc methods'. They permit the operator to see at all times exactly what is going on in the weld pool.

As with most automatic processes where the high welding currents result in large weld pools the continuously covered electrode is used only for welding in the flat and horizontal/vertical positions. It is used for welding longitudinal seams and has a particular application in shipbuilding and structural engineering for fillet welding (pl. 4). For these applications it is notably successful because of its open-arc characteristic and tolerance of poor weather and plate-surface conditions. One version of the process incorporates two welding heads in a portable machine for welding both sides of a T-fillet simultaneously.

Arc-length control. All automatic methods of arc welding in which a consumable electrode is employed must have some form of arc-length control. One of two principles is used, a controlled arc or a self-adjusting arc. For heavy duty automatic machines using relatively thick wires such as the type just discussed, it is general to use the controlled-arc system in which the arc length is controlled by varying the rate at which the electrode wire is fed. The control of feed rate is based on an arc-voltage signal. Higher arc voltages than normal indicate longer arcs than normal and a requirement for an increased feed rate, and vice versa. A great variety of methods are used for varying the wire feed rate according to the arc voltage. The arc-voltage signal may be used to open or close a clutch between the feed rolls and a motor pre-set to run at approximately the right speed. Alternatively, the signal may be used to control, through an electronic circuit or thyratron valves, the armature current of a d.c. shunt motor. Reversible variable speed motors with Ward–Leonard control have of course been used, while yet another method employs two motors connected to the feed rolls through a differential gear. One motor runs at constant speed, the other is varied in speed to produce the required changes in speed of the output shaft and hence in feed rate. A high speed of response by the driving mechanism to changes in the voltage signal is desirable so that wire-feed systems using the controlled arc are designed to have the lowest practicable inertia.

For striking the arc the control system may allow for the electrode wire first to advance to contact the work and then to retract to draw the arc before finally feeding forward again at the required burn-off rate. Another way to strike the arc is to place a small ball of wire wool between the electrode and work to act as a fuse which ionizes the arc gap. To facilitate setting-up, the wire may be advanced or retracted under manual inching control.

Submerged arc

The most universally successful solution to automatic flux-shielded welding is the submerged-arc process, a method developed independently in the U.S.A. and U.S.S.R. in the middle and late 1930s. A bare wire is employed and the flux added in the form of powder which covers completely the weld pool and end of the electrode wire. The arc is therefore submerged completely beneath the flux powder, a fact which contributes both great advantages and some slight disadvantages to the process. It has been widely believed for some years that because molten slags are electrically conducting the heat must be generated in submerged-arc welding by a resistance heating process. Voltage records of the process, however, show features such as short circuits and re-ignition peaks which are characteristic of a true arc.

An advantage of the process is that because the arc is completely enclosed extremely high welding currents can be used without spatter or the entrainment of air. The high current gives deep penetration and the process is thermally efficient because much of the arc is below the plate surface. It is therefore a high dilution process with approximately twice as much plate being melted as wire. Welding currents from 200 to 2000 A are commonly used, although in the early days of the process currents up to 5000 A were employed. For various reasons connected chiefly with the metallurgy of the deposit, these extremely high welding currents are not now generally used as it is preferred to deposit metal in layers to take advantage of the normalizing resulting from reheating (see p. 49).

An open-arc process operated at currents of over 300 A has to be used with care because the arc is an intense source of light with a high infra-red and ultraviolet content. In submerged-arc welding the arc is not seen so that these precautions are unnecessary. By the same token the operator cannot see the weld pool and judge the progress of welding. He must rely on the fit of the joint remaining constant and must either pre-set the traverse of the welding head accurately with respect to the joint or must adjust the position of the head by observing an indicator such as a pointer or a light beam focused just ahead of the powdered flux. With grooved welds roller-guiding devices which run in the prepared edge are often used, and with fillet welds it is often possible to use the vertical plate as a guide to steer the equipment. Experienced users of the processes would probably agree that the inability to see the weld in automatic submerged-arc welding creates more psychological than practical difficulties.

Powdered flux is fused during welding in approximately the same proportion by weight as the electrode wire melted and is left on the weld bead as a layer of glassy slag. Underneath this slag the weld metal has a smooth almost ripple-free surface because of the high heat input which gives a large slowly freezing weld pool in contact with the relatively fluid slag. Submerged-arc welds are remarkable for their neat appearance and smooth contours. Flux not fused during the welding operation is recovered for use again, but care must be taken to see that such flux is not contaminated. Where welding is done on inclined surfaces or near edges, a shelf or some similar device is necessary to support the flux.

Fluxes. Fluxes for submerged-arc welding are granulated to a controlled size and may be of the fused, agglomerated or sintered type.

Originally a fused, crushed and graded flux was used, the advantages claimed for such a flux being that it was totally free of moisture and not hygroscopic. Both the chemical composition and the state of division of the flux have an important bearing on the way it behaves in welding. The approximate proportions of the various principal constituents of fluxes manufactured by a major producer are given in table 4. During the welding process some of the granulated flux is melted to cover both the weld

Table 4.* *Powdered flux and weld metal analysis in submerged-arc welding*

Grade of flux ...	50 (%)	80 (%)	70 (%)	20 (%)
CaO	5	24	28	27
CaF_2	5	5	—	—
MgO	—	12	6·5	7·5
SiO_2	41	38	48	53
Al_2O_3	2·5	13	5	5
MnO	0·75	7·5	10	—
MnO_2	39	—	—	—
Weld metal				
C	0·12	0·11	0·12	0·12
Mn	1·05	1·10	1·00	0·70
Si	0·25	0·30	0·37	0·40
Wire				
C	0·12	—	—	—
Mn	1·80	—	—	—
Si	0·15	—	—	—

* Compiled from published information on the use of Unionmelt fluxes and wires for submerged-arc welding. Figures are approximate and intended to indicate trends. Weld metal analysis is that from all-weld-metal tests. Major constituents only reported.

pool and the metal particles being transferred to the pool from the electrode. In both these stages reactions occur between metal and flux, particularly involving silicon and manganese, elements which play a major role in controlling the strength and soundness of the deposit. The way in which silicon in particular is reduced from the flux is indicated in the lower part of table 4.

The temperature is different in different parts of the arc and pool and the time that a given unit of metal spends in the various stages from incipient droplet on the end of the electrode to its presence in the solidifying pool is influenced by the welding conditions, notably current. The final composition of the weld pool depends, therefore, on both welding conditions and thermo-dynamic considerations. No fully successful quantitative treatment has been derived.

The reduction of silicon from the flux with grade 20 powder gives added deoxidant to help control porosity but leads to a build-up of silicon in the deposit which can adversely affect properties and the risk of cracking in multi-run welds. For this reason it is not used where more than three runs are required. Grade 80 powder, the most widely used, is very fluid when molten and gives an easily removed slag.

Both the composition of the flux and its state of division influence the control of porosity. The submerged-arc process is generally more susceptible than the metal-arc to porosity caused by rusty or dirty surfaces. This is because with an open-arc process water vapour and gaseous products driven off the plate by the heat of welding can escape, whereas in a submerged-arc weld they tend to be retained beneath the blanket of flux. It is for this reason that the fluxes having the highest tolerance to rust and dirt are also those with high permeability, achieved by using a coarse grade with a relatively narrow size range. Where it is required to weld using high currents, however, a closely packed flux is required to give good coverage to the arc. This is arranged by using a finer maximum size of particle and a wider size range to increase the closeness of packing.

Electrode wire. Wire for welding by submerged-arc is used in coil form and is copper coated. This prevents superficial rusting in storage and ensures reliable, low-resistance electrical contact between the welding wire and the copper contacts through which the current is conducted. The diameter of wire used depends chiefly on the welding current required and may range from $\frac{3}{32}$ in. diameter for a current of 150 A to $\frac{3}{8}$ in. diameter for a current up to 3000 A. An indication of the range is given in fig. 11 (p. 31).

Each size of wire can be used over a range of several hundred amperes so that for any welding current there is the possibility of using more than one size of wire. For any given current the smallest diameter of wire will give the deepest penetration. This is because by concentrating the arc root on a smaller wire the arc forces are intensified and metal leaving the end of the electrode does so in a faster moving stream of smaller droplets which scour their way to a greater depth. The weld bead is slightly narrower for a fine wire than for a thick wire at the same current but the main effect of wire size is on penetration.

The composition of wires for submerged-arc welding depends on the material being welded because alloying elements are generally added to the wire and not to the flux as is done with metal-arc welding. With metal-arc welding the ratio of core wire and flux is fixed and hence the weld pool chemistry is to a large extent determined at the time the electrode is manufactured. Variations in technique in submerged-arc welding can alter the ratios of plate, wire and flux melted. Where highly alloyed wires are used, e.g. stainless steels, it may be necessary to add compounds of the alloying element to the flux to slow down metal–slag reactions which would result in losses of the alloying element to the slag.

Arc-length control. As with the open-arc automatic processes it is necessary to have an automatic control over the arc length. For many years the controlled-arc system, in which the wire feed rate is altered to correct for changes in arc length, was used exclusively for submerged-arc welding. It was discovered in Germany during the war, however, that because the electrode wires in submerged-arc welding are generally of small diameter and the current density high there is a natural tendency for the arc length to remain constant even though the wire may be fed at a constant pre-set speed. This ability of an arc to recover its original arc length after a disturbance without a change of the wire feed rate is known as the self-adjusting effect. It is of great importance with the semi-automatic gas-shielded methods of welding to be discussed later. As far as submerged-arc welding is concerned, although the idea of a simple constant speed wire feed motor appears attractive, it has only found wide application in the U.S.S.R. Elsewhere the variable speed controlled-arc system predominates particularly for welding thick plate or for currents above 1000 A where the larger diameters of wire with inherently poor self-adjustment would be used.

With a self-adjusting arc system there are advantages in using a constant voltage or flat characteristic power source (see chapter 4). A power source of this type gives a marked current surge on short circuiting so that arc striking is easy. If a drooping characteristic power source is used, the welding head should be of the controlled-arc type with provision for retracting the wire momentarily after contact is first made with the work so that the arc may be struck. Both a.c. and d.c. power supplies are used, the latter with electrode positive being generally preferred for currents up to about 1000 A because of the good arc stability and deep penetration. Above 1000 A a.c. is preferred, partly because arc blow effects are then less noticeable. Penetration with a.c. is slightly less than with electrode positive d.c. Where very shallow penetration is desired an electrode negative d.c. power supply is used.

Process variables. The important process variables are current, voltage, wire diameter, welding speed and electrode extension. Welding current affects the burn-off rate and penetration directly. Voltage affects the contour of the weld bead and penetration shape. If the voltage is so high that much of the arc is above the plate surface there is a tendency to melt excessive amounts of the flux and the bead becomes wide and flat. Excessively low voltages result in the arc being buried completely below the surface of the plate so that the penetration has a tulip-shaped cross-section. Normal operating voltages for butt welds increase by roughly 6 V per 1000 A as the current is increased, being approximately 35 V at 1000 A as indicated in fig. 17. For fillet welds voltages about 5 V lower are often used to secure a well-shaped fillet.

Welding speed has a marked effect on weld width, welds becoming narrower and more peaky as the speed is increased. Penetration is not affected in proportion as less heat is lost by conduction when the speed is raised so that the process becomes more efficient thermally. Excessive speed results in undercut and ridged or peaky beads.

An increase in electrode extension results in an increased voltage drop between the point of current contact and the arc. The wire is consequently pre-heated so that the melting rate is increased. As would be expected the effect is proportional to I^2R and is, therefore, especially important with high welding currents in thin wires of high resistivity. The melting rate of the electrode comprises that due to the heat generated by the arc at zero electrode extension and that due to the resistance heating effect by the current on its way to the arc through the wire. Wilson, Claussen and Jackson have shown that the arc melting rate itself in terms of mass/min/amp at zero extension is the melting rate at a hypothetical zero wire diameter added to a melting rate proportional to the cross-sectional area of

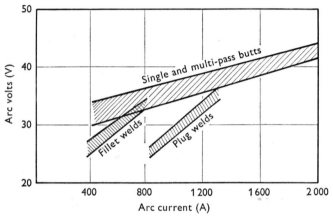

Fig. 17. Variation of operating voltage with welding current for submerged-arc welding.

the electrode. The former is found to be 0·35 lb/min/1000 A and the latter d^2 lb/min/1000 A so that the melting rate at zero extension (in lb/min) is

$$MR_0 = \frac{0 \cdot 35 + d^2}{1\,000}\, I,$$

where d is the wire diameter and I the current in amperes.

The increase in melting rate resulting from electrode extension in mass/min/amp is proportional to the product of current density and electrode extension, the total melting rate, MR (in lb/min) being

$$\frac{I}{1\,000}\left[0 \cdot 35 + d^2 + 2 \cdot 08 \times 10^{-7} \times \frac{(IL)^{1 \cdot 22}}{d^2}\right],$$

where L is the extension in inches.

An increase in melting rate of 45 per cent results from increasing the electrode extension from 2 to 6 in. with a $\frac{3}{16}$ in. diameter steel wire carrying 600 A.

Applications. Submerged-arc welding has found its main application with mild and low alloy steels, although the process has also been used for copper, aluminium and titanium base alloys with the development of suitable fluxes. It is a method used mainly for the downhand welding of thicknesses above $\frac{3}{16}$ in. where the welds are long and straight. The welding head itself may be moved over the work on a self-propelled carriage or gantry or the work may be revolved beneath the stationary welding head. The method is widely used for both butt welds and fillet welds in shipbuilding, pressure vessel, structural engineering, pipe welding and storage tank fabricating industries. For the latter purpose special self-propelled machines with devices for supporting the flux are used for welding the girth seams in position on-site.

To give flexibility to the process for shorter more awkward joints a manual version has been developed in which the contact tube and a portable flux dispenser held in the hand are connected by a flexible hose to the

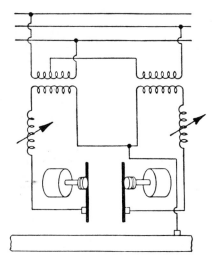

Fig. 18. Two electrode 2-phase (Scott) power connection.

main wire feed unit. Because the arc cannot be seen this device is difficult to use without aids to help the welder guide the welding head over the work and maintain a constant speed. Semi-automatic CO_2 welding (to be discussed in the next chapter) is generally a better solution to the problem of providing continuous high current welding with flexibility.

Multiple electrodes. Although submerged-arc welding has been discussed as a single wire process there are several modifications involving the use of more than one wire which extend the range of the process. These techniques are known collectively as multiple electrode welding. In parallel electrode welding two electrodes are usually spaced between $\frac{1}{4}$ and $\frac{1}{2}$ in. apart and are connected to the same power source. If the electrodes are used in tandem, one following the other, an increase in welding speed of up to 50 per cent may be obtained. When used side by side the arrangement permits wider grooves to be filled. A.c. is preferred because multiple d.c. arcs of the same polarity tend to pull together. If the wires are supplied with power from separate power sources the technique, indicated in fig. 18, is known as multiple-power tandem welding or multiple-power multiple-wire welding. To prevent arc blow effects at least one of the arcs should be a.c. when high currents are used and electrode separation is 1–2 in. greater than for the parallel arrangement. Even higher welding speeds can be achieved with multiple-power than with parallel electrode welding. Where high deposition rates with little penetration are required, as in surfacing, an arrangement called series welding, shown diagrammatically in fig. 19, is sometimes employed. Two welding heads are used, a single power source

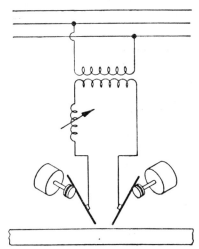

Fig. 19. Multiple electrode submerged-arc, series connection.

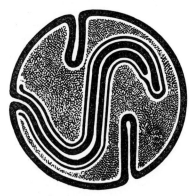

Fig. 20. Cross-section of a flux-cored electrode produced by enclosing flux in a folded strip which is subsequently drawn.

being connected across them so that two independent arcs in series are drawn. Each arc has a different polarity so that they tend to spread away from each other.

Cored electrode-arc welding

Another solution to the problem of introducing current to a continuous fluxed wire is to put the flux into tubular steel electrodes drawn from strip or enclosed in a folded metal strip of the type shown in fig. 20. Both methods permit a wide variety of special wires to be produced easily. Cored electrodes are generally used with an auxiliary shielding gas such as carbon dioxide and where this is done the flux core is there to provide the essential deoxidants, silicon and manganese and any alloy additions required. Arc stabilizers can also be included. Used in this way the flux/gas method resembles gas-shielded welding in that semi-automatic operation is feasible. Because of the necessity to enclose the flux however the electrode sizes tend to be on the upper limit normally used for gas-shielded welding. The process is not quite so clean as the gas-shielded method but the presence of an appreciable layer of slag on the weld gives improved weld shapes in the horizontal–vertical position and reliable shielding.

If an auxiliary gas shield is not used the flux in the core must contain, besides the deoxidants, compounds that decompose in the arc to give gaseous or volatile products which can sweep the air from the arc and weld pool region. Calcium carbonate is one such material.

Magnetic flux/gas-arc welding

A method which overcomes the disadvantages of the bulk of the cored electrode wire and the lack of visibility of the submerged-arc process uses a magnetizable flux. This is carried in a stream of carbon dioxide which emerges through a nozzle surrounding the bare electrode wire. Because of the magnetic field surrounding the electrode wire when current is passing, the finely powdered flux is attracted to and adheres to the wire, forming a coating. Appreciable quantities of flux may be carried on the wire, the normal flux : electrode wire ratio being between 1 : 3 and 1 : 2 by weight. This ratio is pre-set and maintained by a metering device coupled to the feed rolls. The carbon dioxide provides auxiliary shielding. Equipment for welding with a magnetizable flux is similar to that for gas-shielded welding and the wire sizes used are in the same range.

Magnetic flux/gas-arc welding is suitable for making joints in mild steel in the flat and horizontal–vertical positions, in the latter of which the flux helps to maintain the weld shape. Although currents as low as 100 A have been used the main application of the process is in the range 250–500 A, which is above that where the dip transfer CO_2 arc method operates.

Carbon-arc welding

Although welding with a carbon electrode was one of the earliest processes used carbon-arc welding is now rarely employed. The process was used in two forms, a direct arc between the carbon electrode and the work, and the indirect method with the arc between two carbon electrodes. With the direct arc the electrode was surrounded with a coil carrying the welding current so that the arc was constricted electro-magnetically and its axial stiffness improved. Originally the process was used without shielding, but fluxes were later introduced for welding both steel and copper base alloys. The carbon arc was used for welding sheet steel, copper and even aluminium and also for brazing. Its chief use now is as an arc-gouging process. A carbon-arc—usually d.c. unless cored electrodes are employed—is used to melt small areas of the metal to be gouged and the molten metal is then blasted away by compressed air issuing through a jet in the electrode holder.

EDGE PREPARATION AND PENETRATION CONTROL FOR ARC WELDING

Reasons for edge preparation

One of the most important welding variables is edge preparation. The reasons why edges are prepared are similar whatever fusion welding process is used, and the choice of design is usually fundamental to the production of a successful joint. Why are edges prepared and masses of metal hewn out of thick joints to be replaced piecemeal by individual runs of weld metal? To understand the reasons it is necessary to consider what happens as the thickness to be welded is increased.

When thin metal is welded it is possible to use the simplest of all joint designs—the square butt joint. The metal is penetrated completely and the penetration bead is held in position by surface tension. As the metal thickness is increased the welding current must be raised to ensure penetration. Two things then occur; either the weld bead width increases in keeping with the penetration causing the whole pool to become larger or there is an increase in arc force so that the weld metal cannot be contained.

The forces acting downward on a molten weld pool derive from the mass of the liquid metal, the kinetic energy of transferring metal particles and the arc force as a result of plasma streaming. If these forces exceed the surface tension acting on the underside of the weld pool, the weld pool drops through the joint. The mass of liquid metal depends on the size of the weld pool and therefore on current and welding speed. Arc force is controlled chiefly by current. Surface tension can be modified to some extent by the presence of fluxes and slags.

Backing techniques

The difficulty of supporting the weld bead can be overcome by using a backing technique. Several such techniques are available:

(1) *Permanent or fusible backing*, in which a strip of the parent metal is inserted underneath the joint and is fused to become part of the joint.

(2) *Temporary backing bars* in which the weld bead is supported but no fusion occurs to the backing bar. Such backing bars are often made of copper so that the molten metal is chilled rapidly. A proprietary make of ceramic-coated steel strip can also be used effectively with gas-shielded welding. To help secure thorough fusion of the root it is often helpful to allow the root bead to drop a fraction of an inch so that oxide films are broken. Backing bars are, therefore, grooved with rounded sides so that a suitable shape is imparted to the underbead. A variation of this technique employs a spring-loaded backing shoe which is moved along under the molten weld pool.

(3) *Flux backing*. In this technique powdered flux is supported in a

trough underneath the weld and is pushed against the joint by such means as an inflatable hose. This method is more suitable for long lengths and more tolerant of variations in fit-up than the backing bar of copper.

(4) *Root backing.* The commonest method of supporting a weld bead is to arrange for the penetration to be incomplete so that the unmelted parent metal provides support. The joint must then be completed with a weld from the other side, often after a groove has been cut, a process known as back-gouging. Alternatively, the process may be reversed by using a weld bead in the root called a backing-run which supports a larger weld bead laid subsequently. For example, a metal-arc weld may be laid in the root of a joint to support a submerged-arc weld.

(5) *Gas backing.* With metals which are readily contaminated by contact with the atmosphere or which form refractory oxides when hot, it is necessary to protect the heated underside of the weld. This may be done by the gas-backing technique in which an inert gas is directed against the penetration bead. Two methods are used—tubes or cylinders may be sealed and filled with the inert gas, or ducts may be cut in a backing bar through which the gas is fed directly to the penetration bead. By preventing heavy, uneven oxidation gas backing can improve the shape of the penetration bead. As the gas has only a marginal effect in supporting the weld bead gas backing cannot be classed with the other methods of supporting the weld bead.

The use of backing techniques allows high welding currents, the use of automatic welding, large weld pools, minimum edge preparation and gives good tolerance to fit-up. It is also possible to increase welding currents to weld thicker material in a single pass, but if this policy is pursued too far fresh limitations begin to take effect.

There are three such limitations: (*a*) the weld bead may become so large and cool so slowly that an unfavourable microstructure results with inferior properties, or (*b*) the metal transferred from the electrode piles up to form an excessive reinforcing bead, or (*c*) the weld bead becomes too deep in proportion to its width so that centre-line cracking results on solidification. The penetration ratio of depth to width should not exceed 3:2.

Prepared joints

A further increase in thickness welded requires that metal is removed from the joint edge by grooving or bevelling. This accommodates the unwanted extra metal from the electrode because of the dependence of electrode melting rate on current, but it also allows multi-run welding and brings thick metal within the range of low current welding processes. In multi-run steel welds it is usually considered an advantage that each weld bead reheats the one deposited previously allowing the relatively coarse, directional grain structure to recrystallize giving a fine-grained structure.

4

With processes in which flux is used the groove cannot have steep sides or slag would be trapped in the undercut where the edge of the weld melts into the groove. Clearly the groove must have sloping sides so that slag can be removed between passes. Even with those gas-shielded processes using argon in which there is negligible slag, however, a sloping side to the groove is desirable since undercut areas are shielded from the arc cleaning action and oxide inclusions or lack of side fusion can result. A groove with sloping sides also allows the welder to see and manipulate his electrode or torch in the root of the groove.

Although a grooved joint and multi-run welding is a procedure which ought to be applicable to metal of any thickness, factors such as economics, the risk of weld defects, the need for pre-heating, residual stresses and distortion have greater effect as the thickness and number of weld runs is increased. Obviously the thicker the work becomes the greater is the volume of metal which is removed in the edge preparation and the more wasteful the process. On very thick joints less metal is removed with a U preparation than a V and although expensive to machine there comes a point where the extra cost of the U preparation is balanced by the saving in metal deposited. A double-sided preparation also reduces the volume of weld metal required as well as reducing distortion because weld runs can be deposited on alternate sides to balance shrinkage stresses. The chief difference between edge preparations for manual and automatic welding is that close square butt joints can be used to greater thicknesses and the root face can be thicker with automatic welding. This results from the ability to use higher currents giving deeper penetration and fewer runs and is one reason why welds made automatically can be more economical than the same welds made manually. Automatic welding, however, requires greater uniformity and accuracy of edge preparation than manual welding.

Controlled penetration techniques

Although the backing techniques described can be used in a joint prepared with a groove it is also possible to employ controlled penetration techniques. With this method a small unsupported root run is made first followed by runs which are called filling and capping runs. Controlled penetration techniques are widely used where access to the underside of the joint is restricted so that a backing bar cannot be employed, for example with pipes.

It is necessary to have an edge preparation with a wide V or U so that a small, low current, easily controlled weld bead may be made in the root. Welding processes such as oxy-acetylene or tungsten-arc are often used for fusing the root as they allow good control of the weld bead and permit the separate supply of heat and filler metal. Where the shape of the underbead is especially important and access is possible from only one side the root insert technique can be used. A specially shaped rod is positioned in the root of a single V joint and is fused into the joint by the tungsten-arc process. The shape of the underbead is also improved when an inert gas is used to shield the underbead.

Because the weld bead with controlled penetration techniques is supported entirely by surface tension there is not much latitude with welding conditions and the welding speeds achieved are appreciably less than when backing techniques are used.

ELECTRO-SLAG WELDING

As the thickness of the metal welded increases, multi-run techniques become uneconomical. The use of automatic welding, however, with high current large passes in the flat position can give a weld pool so large that it runs ahead of the electrode out of control resulting in inadequate fusion. The difficulties of heavy pass welding can be overcome by turning the plates into the vertical position and arranging a gap between them so that the welding process becomes rather like continuous casting. Weld pools of almost any size can be handled in this way provided that there is adequate energy input and some form of water-cooled dam to close the gap between the plates and prevent the weld pool and slag running away. Vertical welding is an essential feature of the electro-slag welding process which was developed from the long-established submerged-arc welding process by the Paton Institute of Kiev in about 1953. Since then the process has been widely adopted in the U.S.S.R. and Europe for the welding of heavy steel sections. Three major variations of the process exist and are known as wire electrode, plate electrode, or consumable guide according to the type of electrode employed.

Figure 21 shows the basic arrangement for the most common, wire electrode type, electro-slag welding and the way in which a bath of slag is carried on top of the molten metal pool. The electrode wire is introduced through this slag pool where the heat necessary to maintain the process is developed. The process is called electro-slag welding because when the slag is molten it is electrically conducting and heat is generated by the resistance offered to the current during its passage from electrode wire through the slag into the weld pool. Although thin plates, less than 1 in. thick, can be welded with the electrode wire positioned accurately and equally between the faces of the plates to be welded, it is necessary with thick material to traverse the electrode to and fro so that uniform fusion of the joint faces through the thickness of the joint can be obtained. With material over 4 in. thick more than one electrode wire may be employed. As the electrode wire is melted away the level of the weld pool rises and the water-cooled shoes are raised at the appropriate rate. Speeds of several feet per hour are achieved.

The molten slag bath over the weld pool is both heat source and shielding agent. Most fused salts become increasingly electrically conducting as their temperature is raised. Once the bath of molten slag has been established over the weld the process proceeds by resistance heating, the main current path being from the end of the wire to the pool immediately underneath. As with arc welding where the current carrying arc plasma streams, or flows, in electro-slag welding the slag tends to be set in motion as

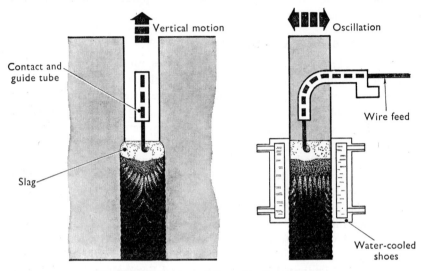

Fig. 21. Electro-slag welding with wire electrode.

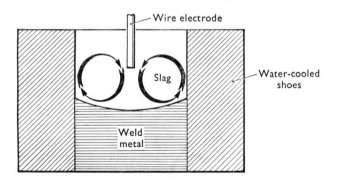

Fig. 22. Circulation of slag resulting from the flow of welding current.

shown in fig. 22 causing turbulence in the slag bath and a shallow crater under the electrode. The molten slag with a temperature up to 2000 °C, washes the joint edges, melting into them to produce a weld with a dilution, high by arc-welding standards, of up to about 50 per cent.

This high dilution results from the use of square edge preparations and gaps of between 1 and 1½ in. for plate of 1 in. thickness upwards. Special techniques have been developed for electro-slag welding thin plate, ¼–1 in.

thickness, however, in which the dilution has been reduced to about 25 per cent by using high conductivity, low viscosity slags.

Starting procedure. The process is only stable when a slag bath and weld pool of adequate size and heat content have been established. At the beginning of each weld, therefore, a U-shaped run-on plate is provided. Welding is begun by striking an arc from the electrode wire to the root of the run-on plate with a small quantity of flux. As the temperature builds up more flux is added and the molten slag periodically quenches the arc, which is re-ignited as the wire short circuits to the plate. After a series of such flashes current conduction begins through the slag and the process is under way. At the start, conditions must favour an arc, but once the electro-slag process is working arcing is to be avoided and the welding conditions and power supply are deliberately chosen to be unsuitable for maintaining an arc. A.c. is the preferred power supply for thick plate, although d.c. is used for plate less than 1 in. thick. The equipment frequently has tappings which allow the high open-circuit voltage necessary for the arc start to be reduced once the electro-slag process begins. The main factor influencing the change from arc to electro-slag operation, however, is the depth of the slag pool. An adequate depth of molten slag ensures that arcs are quenched and that current conduction is by electric resistance.

Power supplies and controls. The power source for electro-slag welding usually has a flat or near-flat output characteristic. This is partly a result of the low open-circuit voltage requirement, but it means in practice that a self-adjusting control system can be used in which pre-set constant-speed wire-feed motors are employed. The system is analogous to that of the self-adjusting arc used in gas-shielded welding, except that the resistance of the molten slag path replaces the resistance of the arc. Operating voltages of about 50 V are usual with thick plate and are determined by the open-circuit voltage. Weld pool depth increases slightly with increase in voltage but the main effect is on weld width (see fig. 23), and consequently on dilution which increases considerably with increase in voltage. When welding plate of less than 1 in. thickness the voltage may be reduced by about 15 V.

Welding current for a flat characteristic power source is determined by the electrode feed rate, there being a linear relationship between electrode feed rate and current as in arc welding. For any given feed rate, however, between 150 and 200 A less current is passed compared with arc welding. Because the joint is almost exactly filled with filler metal the electrode feed rate is also directly related to welding speed. Weld pool depth also varies in a linear manner with electrode feed rate, although the weld width at first increases with increase in feed rate and then decreases. This is because of low total power at low feed rates and reduced power per unit length of weld at high feed rates and currents. For an average size wire of $\frac{1}{8}$ in. diameter welding currents would be between 300 and 600 A.

Although the rate of vertical travel or welding speed is directly related to the rate at which the joint is filled up by molten electrode metal, this speed cannot be pre-set at a constant value because of minor variations in other parameters such as joint fit-up. Although many users prefer manual control several systems are available for sensing automatically the level of the weld metal and raising the copper shoes accordingly. Two widely used methods employ temperature measurement in one of the shoes or voltage pick-up from a probe in one of the shoes. The latter is the preferred method because of its simplicity. When the probe is in contact with molten slag,

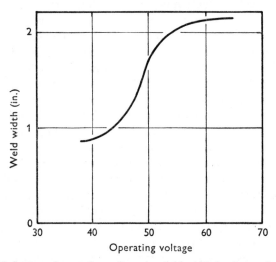

Fig. 23. Influence of operating voltage on weld width in electro-slag welding.

which is electrically conducting, a voltage can be picked up between the electrode guide and the probe. If the level of the metal rises too high the probe is covered not by molten slag but by solidified slag and solid metal so that contact is lost. The loss of signal operates the control system which then raises the shoes until contact is restored.

The depth of the slag bath is checked by dipping a wire into the pool. Wastage of slag occurs by leakage between shoes and weld surface and fresh flux must be added by the operator from time to time to maintain a constant slag level. Slag pool depths of $1\frac{1}{2}$–$2\frac{1}{2}$ in. are used, the latter for thinner material.

Flux and wire. Slags must have not only the right electrical properties they must also have suitable viscosity. This must not be so low that excessive leakage occurs between shoe and plate or so high that the weld surface does not have a smooth form. A highly adherent slag is also to be avoided. Generally the fluxes used are similar to those used for submerged-arc welding but they contain more fluorides. A typical flux for welding carbon steels would have the analysis for main constituents given in table 5.

Table 5. *Electro-slag welding flux for mild steel*

	%		%
CaO	4–7	SiO_2	33–36
CaF_2	13–19	Al_2O_3	11–15
MgO	5–7	MnO_2	21–25

Additions of fluoride markedly reduce the viscosity and improve the electrical conductivity of the molten slag.

Reactions occur between the constituents of the slag and the electrode and weld metal, the most important of which concern manganese and silicon. As in other forms of flux-shielded welding MnO and SiO_2 are reduced by iron resulting in build-up of FeO in the slag. Besides slowing up the reduction process and therefore giving a weld deposit lower in Mn and Si this causes surface arcing in the slag bath. Reduction of MnO and SiO_2 may be considerable in the first 4 or 5 in. of weld but thereafter occurs more slowly. The high fluoride content of the fluxes and the protracted thermal cycle result in losses of silicon as the volatile silicon tetrafluoride. As with arc welding, changes in welding parameters which increase pool temperatures or the time the electrode wire takes in transfer tend to increase the rate of these chemical reactions.

Electrode wires are chosen to suit the composition of the parent metal and are similar to those used for submerged-arc welding. Flux-cored types of wire are used as well. Generally these wires introduce manganese and silicon to the weld to prevent the reaction $C + FeO \rightarrow Fe + CO$, which gives porosity. With the electro-slag process, however, there exists the possibility of changing the weld analysis by adding special alloys in various ways, such as separate wire feed, powder, or rods or plates placed in the gap.

Electrode oscillation. The welding variables already mentioned are not altered greatly to accommodate changes in the thickness being welded. Above 4 in. thickness, however, it is usual to employ more than one electrode wire. Oscillation of the electrode though the thickness of the joint is necessary to give a weld of uniform width. To prevent the weld from narrowing at the surfaces where heat is lost to the water-cooled shoes the electrode is allowed to remain stationary for a brief period at the point of reversal. When more than one wire is used the amplitude of oscillation is made less than the distance between the electrodes so that the stationary period for the inner electrode does not fall at the same position on the cross-section each time. If this occurs, excessive heating and an increase in weld width will result at this position. Speed of oscillation is not critical. Weld cross-sections are often slightly barrel-shaped, but for the purposes of calculating weld dilution may be assumed to be rectangular. The proportions of parent-metal and electrode metal in the final weld may be calculated from the ratio of weld width and gap width.

Weld structure. Because of the large slowly moving weld pools electro-slag welds are characterized by protracted thermal cycles. Unlike other methods of welding the joints are always completed in a single pass so that there is no opportunity to exploit the reheating effect of subsequent runs to produce a refined grain structure. The difference in thermal history between a multi-pass manual metal-arc weld in 3 in. thick mild steel plate and an electro-slag weld in a similar thickness is shown in fig. 24. Tempera-

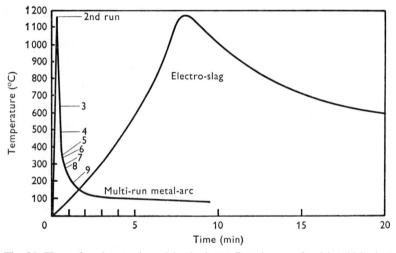

Fig. 24. Thermal cycle experienced in the heat-affected zone of a 3 in. thick electro-slag welded plate, compared with multi-run metal arc where figures indicate peak temperatures at root for successive runs.

ture measurement was recorded at the centre of the thickness and because the metal-arc weld was made with a double V technique the series of temperature peaks resulting from each pass is repeated. As the weld is filled up the peak temperatures become lower.

The protracted thermal cycle for electro-slag welding results in a weld structure of large grains with a marked tendency to columnar growth. Both the weld metal and the heat-affected zone can have inferior impact properties to the parent metal, and, while a fine-grained structure with acceptable impact properties can be achieved by normalizing, many joints are used in the as-welded condition. The cleanliness of the deposit and the low incidence of weld defects are regarded as advantages. The less rapid thermal cycle of the electro-slag process makes it capable of welding safely thick steels with a hardenability dangerously high for other welding processes.

A defect which is characteristic of electro-slag welds results from the unfavourable orientation of the columnar grains and is the main factor limiting welding speed. At slow speeds the weld pool is shallow and crystal growth taking place normally to the solidification front is predominantly in the direction of welding. With an increase in speed the pool becomes deeper and crystal growth tends to be from the joint faces and centre-line cracking can result. The tendency to cracking depends, therefore, on two

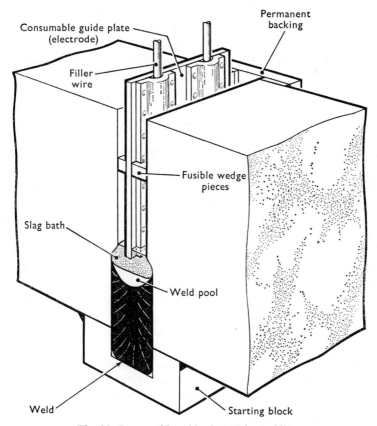

Fig. 25. Consumable guide electro-slag welding.

factors; the inherent crack sensitivity of the material and the width to depth ratio of the weld pool. For mild steels the weld width should be at least half as much again as the depth, average values for the ratio being 1·5–4·0 which means that the maximum welding speed for plate over 3 in. thick is about 3 ft/h.

Consumable guide electro-slag welding

This variation, illustrated in fig. 25, also employs an electrically conducting slag and is used for vertical welding.

The equipment for welding is considerably simpler than that for the wire electrode type of electro-slag welding chiefly because the welding head and wire feed mechanism do not need to be moved up the joint as the weld is made. After the joint has been assembled for welding with a gap slightly narrower than for wire electrode electro-slag welding a grooved steel guide plate is fitted into the joint. The essential feature of this plate is that it should contain passageways down which wire can be fed, a requirement that can be met in many ways from very thick-walled tube to a plate assembled from steel bars and sheet metal. If an alloyed weld metal is to be deposited alloy strips can be incorporated in the plate, but allowance must be made for dilution. The guide plate is connected to the power source and must be insulated from the joint faces by blocks or wedges of wood or asbestos compound which may be removed progressively as the joint is completed.

A distinctive feature of consumable guide electro-slag welding is that because there is no wire-feed contact tube to be moved up the joint it is possible to weld joints where access is available from one side only or indeed where there is a permanent backing bar on both sides of the joint. For making butt joints a full-length backing bar may be fastened on one side of the joint and short lengths, 12–18 in., of backing bar used on the other side. Normally two short lengths held by magnetic clamps are 'leap-frogged' up the joint so that there is always access to the consumable guide to within about 12 in. of the weld pool. Butt joints in plates from $\frac{3}{4}$ to 3 in. thick can be made in this way in lengths up to at least 30 ft. Other joints such as full penetration T joints which are difficult to make by the wire electrode method because of the restricted access can be made easily by consumable guide methods, also curved joints and those between shaped pieces where the joint cross-section is not constant.

The electro-slag process is started in the normal way and as the weld pool rises due to the metal added from the electrode wires the guide plate is melted away. Welding speed in this type of electro-slag welding depends on the rate at which the electrode metal fills the gap between the stationary guide plate and the work, and the welding speed can be considerably higher than for the wire electrode process.

Plate electrode electro-slag welding

Another type of electro-slag welding in which plate or bar electrodes are suspended in the joint gap and lowered slowly as they are melted away by the slag bath. A three-phase system may be used with three plate electrodes connected in star fashion. The process is controlled by setting the feed motor to lower the plates at approximately the correct rate, final control being through a start-stop mechanism operated by the working voltage.

Applications

Electro-slag welding is primarily a method for welding heavy thicknesses in the vertical or near vertical position. For metal over 2 in. thick, where it competes mainly with submerged-arc welding, it has the advantage of speed and reduced distortion. Because the joint is completed at one attempt and not in a series of passes rotational distortion is absent. The mechanical properties of electro-slag welds may be inferior to multi-pass submerged-arc welds, however. Electro-slag welding is used in pressure-vessel construction, heavy structural engineering, press frame and water turbine fabrication. The wire electrode type is most common and the equipment comprises a vertical boom up which the electrode feed and oscillation mechanisms and the water-cooled shoe supports are moved by rack and pinion or chain drive. Consumable guide electro-slag welding is particularly useful for the shorter lengths of weld where a variable cross-section is required. The plate-type electrode is used for short welds of exceptional thickness, where the joint cross-section approaches a square.

With material thinner than 1 in. electro-slag welding would only be used if it was essential that the welding had to be done in the vertical position. The process shows a distinct economic advantage over manual metal-arc welding of long vertical seams of above $\frac{1}{4}$ in. thickness. Suitable applications for thin plate electro-slag welding occur in the welding of ships, storage tanks and heavy structural engineering. For this type of application light-weight portable equipment has been developed which can walk up the steel plates using magnetic shoes. Although electro-slag welding is mainly used for mild steel and low alloy steels it is capable of being used for high alloy, stainless steels and titanium. The slag cover is not impervious to gases and both oxygen and hydrogen may pass through the slag to the weld metal. For this reason additional inert-gas shielding is employed when the electro-slag process is used for welding reactive metals such as titanium. A special oxide-free flux is also used.

In various forms, additional to those described, the electro-slag process is used for such widely different purposes as surfacing, ingot refining, ingot hot-topping and the butt joining of bar.

ELECTRO-GAS WELDING

The technique of vertical welding between water-cooled copper shoes characteristic of electro-slag welding is also used in a superficially similar process called electro-gas welding. In electro-gas welding, however, heat generation is by an electric arc which is struck from a flux-cored electrode to the molten weld pool. The flux from the flux-cored electrode forms a thin protective layer but does not give a deep-slag bath as in electro-slag welding. Additional shielding of the weld pool is provided by a shield of carbon dioxide or argon-rich gas which is fed over the weld pool through the top of each copper shoe. Mechanically the apparatus is similar to that for the wire electrode type of electro-slag welding.

Electro-gas welding can be used on thicknesses from $\frac{1}{2}$ to 3 in., although it is chiefly used at the lower end of this range, for example in shipbuilding and the site fabrication of storage tanks. Because it is an arc-welding method and capable of speeds in excess of electro-slag welding for comparable thicknesses, the thermal effects and therefore the metallurgy in the heat-affected zone more closely resembles a submerged-arc weld than an electro-slag weld. For the same reason an electro-gas weld can be started without the necessity of a starting block, and if for any reason a weld run is stopped it can be re-started with less difficulty than an electro-slag weld.

Bibliography

BEAMA. *Guide to Covered Electrodes.*

Belton, A. R., Moore, J. J. and Tankins, E. S. (1963). Slag-metal reactions in submerged-arc welding. *Weld. Res.* (Suppl. 28), **7**, 289–97.

Birkbeck, E. J. (1946). Firecracker production welding. *Welding*, **14**, 396–402.

Chouinard, A. F. and Monroe, R. P. (1957). A new CO_2 welding process (cored electrode). *Weld. J.* **36**, no. 11, 1069–73.

Davis, N. and Telford, R. T. (1957). Manual magnetic-flux gas shielded arc welding of mild steel. *Weld. J.* **36**, no. 5, 475–80.

Fieldhouse, A. B. (1959). Some facts about iron powder electrodes. *Welding and Met. Fab.* **27**, no. 2, 61–70.

Flintham, E. (1954). The development of metal-arc welding and electrodes during the past half century. *Br. Weld. J.* **1**, no. 1, 3–10.

Gledhill, P. K. (1947). Viscosity of welding slags. *Metallurgy of Steel Welding.* B.W.R.A.

Jackson, C. E. and Shrubsall, A. E. (1953). Control of penetration and melting ratio with welding technique. *Weld. J.* **34**, no. 4, 172S–8S.

Kerr, J. G. and Elmer, D. A. (1952). Chromium recovery during submerged-arc welding. *Weld. J.* **31**, no. 9, 431S–8S.

Kubli, R. A. and Shrubsall, H. I. (1956). Multi-power submerged arc welding of pressure vessels and pipe. *Weld. J.* **35**, no. 11, 1128–36.

Latimer, J. (1952). Automatic welding. *Trans. Inst. Weld.* **15**, no. 3, 71–85.

Leder, P. J. (1957). The application of CO_2 shielding to the continuous covered electrode process. *Brit. Weld. J.* **4**, no. 5, 274–81.

Lucey, J. A. (1958). Practical aspects of automatic welding. *Weld. Met. Fab.* **26**, no. 10, 375–80, 397–402.

Mathias, D. L. (1958). The use of iron-powder electrodes in the U.S.A. *Industri Information*, no. 6. Hoganas: Billesholms AB.

Norcross, E. J. (1965). Electro-slag and electro-gas welding in the Free World. *Weld. J.* **44**, no. 3, 177–86.

Paton, B. E. (1962). *Electro-Slag Welding* (English Translation of Russian text). American Welding Society.

Pocock, P. L. (1945). Carbon-arc welding. *Welding*, **13**, 54–60, 154–63.

Recommended Joint Preparations for Fusion Welding of Steel (I.I.W. Document) (1962). *Brit. Weld. J.* **9**, no. 1, 13–28.

Taylor, H. G. (1956). Modern welding. *Royal Soc. Arts*, **104**, no. 4983.

Warren, D. and Stout, R. D. (1952). Porosity in mild steel weld metal. *Weld. J.* **31**, no. 8, 381S–92S; **31**, no. 9, 406S–20S.

4

GAS-SHIELDED ARC WELDING

HISTORICAL BACKGROUND

The idea of using a gaseous shielding medium to protect both the electric arc and weld metal from contamination by the atmosphere is almost as old as the covered electrode. Roberts and van Nuys in 1919, and others several years later, considered the problem and a variety of gases were proposed from the inert gases to hydrogen and hydrocarbons. In the 1930s the interest began to centre on the inert gases but it was not until 1940 that experiments were begun at the Northrop Aircraft Co. of U.S.A. with the deliberate intention of developing a practical inert-gas welding method. The metal to be welded was melted by an arc struck from a tungsten electrode, in an atmosphere of the inert mon-atomic gas helium.

The original apparatus comprised the simple tungsten electrode torch and a d.c. generator. Arc starting was by brushing the electrode on the work but this led to contamination of the electrode and a high-frequency spark generator was added to the equipment so that an arc could be struck from the electrode without touching it on the work. At first both electrode negative and electrode positive polarity were used, although the negative polarity was favoured because less heat was generated at the tungsten electrode, which remained relatively cool.

With the desire to weld thicker material welding currents were increased to over 100 A and it was no longer possible to use the electrode positive polarity because the tungsten electrode became so hot that molten tungsten dropped off into the weld pool. The higher welding currents also necessitated water cooling of the body of the torch because of the increased amount of heat conducted back along the electrode.

By 1944 it was recognized that electrode polarity was of greater significance than had appeared at first. Up to that time the inert-gas arc process had been used principally on thin-gauge magnesium and stainless steel, but attempts had also been made to weld aluminium with which it was found necessary to employ a flux. It was observed, however, that oxide removal could be accomplished by the arc itself on electrode positive d.c. or in a.c. welding, thus making a flux unnecessary. Unless a certain minimum open-circuit voltage was available when welding aluminium with a.c. the oxide film was not broken down so that the a.c. was rectified and welding was impossible. By 1946, however, it had been found that the spark ionizer could be made to stabilize the a.c. arc. Gradually a preference emerged for argon over helium in manual welding, largely as a result of the smaller change in arc voltage with arc length when welding with argon. This made the process less critical from the welder's point of view.

Once a start had been made to the welding of aluminium by the inert-gas tungsten-arc method there began a period of rapid development because of the new range of applications opened up. Although limited for several years to welding sheet material at less than 150 A there was now a demand to go to higher currents. Metal gas nozzles were replaced by ceramic ones, these in turn being replaced by water-cooled metal nozzles when it was found that the ceramic nozzles had a limited life. The water-cooled torch body and power lead was now essential to give lightness and flexibility to the torch, and because the high-frequency ionizer was left on continuously great attention had to be paid to insulation.

Although the high-frequency ionizer stabilized the arc it did not affect the inherent unbalance between the voltage on alternate half-cycles which resulted in a d.c. component that tended to saturate the transformer. At first this was overcome by applying a similar d.c. voltage, but of opposite polarity, to the circuit so that the d.c. component was balanced out. This was done with storage batteries, but subsequently it was found that large capacitors in series with the arc had the same effect.

The purity of the shielding gas was improved from 98 to over 99·95 per cent as the process developed, particularly as a result of the need for high purity gases for the welding of aluminium alloys and reactive metals. Argon, the only inert gas available outside the U.S.A., gained in popularity even in that country, being the chief gas used for manual welding, although the higher arc voltage and, therefore, greater penetration of the helium-shielded arc was found of value in automatic welding. Both helium tungsten-arc and argon tungsten-arc techniques were rapidly applied to the welding of a range of non-ferrous metals which had proved difficult to weld by other methods.

With aluminium, as with magnesium, the new process gave greater scope to the engineer because of the absence of flux. Previously fillet welds and other types of joints in which flux might be trapped had to be avoided because of the danger of corrosion after welding. The more concentrated heat input of the tungsten-arc welding process over gas-welding enabled welding speeds to be increased and improved the metallurgical quality of welds. Although there were many advantages to the process it was also found to have limitations. The separate addition of filler metal required the use of both the welder's hands, access to difficult joints tended to be restricted and positional welding was slow and difficult.

In 1948, however, the second important gas-shielded process made its appearance and was to prove capable of being used satisfactorily on many of the types of joint which were not ideally suited to the tungsten-arc method. In tungsten-arc welding the electrode was non-consumable, but in the new method the electrode was in the form of wire which was consumed during welding to provide filler metal for the weld. This wire was fed from a coil to the arc at the same rate as it was melted away. The term metal-arc is used to describe an arc-welding process in which the electrode is consumed during welding to provide filler metal for the weld and the new process therefore became known as inert-gas metal-arc welding. It was not long before gases other than inert were used so that the process should now strictly be described as argon metal-arc, helium metal-arc or CO_2 metal-arc, etc., as appropriate, with the general title of gas metal-arc for the whole series.

In the first apparatus the wire was pushed through a flexible tube to a pistol-type torch where contact was made with the welding current conductor. Argon gas to shield the weld pool was passed through a nozzle surrounding the filler wire. Although the torch was held in the hand the process possessed certain characteristics usually associated with automatic welding. It was the first manual process to utilize the principle of the self-adjusting arc in which the arc length is held constant during welding, irrespective of movement by the operator. An essential feature of the process, which made it possible to use both a self-adjusting arc and the flexible feed tube to the torch, was the small diameter of the electrode wire, usually about $\frac{1}{16}$ in. Metal was transferred axially from this wire electrode to the work in a stream of fine drops.

Development of the inert-gas metal-arc method in the early 1950s was closely associated with the welding of aluminium alloys which at that time were becoming established as structural materials, in particular for shipbuilding where a process was needed which would weld in any position. Had the need for structures in aluminium alloys not existed in 1950 the process might well have been developed more slowly, and it was fortunate that aluminium was one of the first metals to be tried for, as is now known, metal is transferred across the arc more satisfactorily with aluminium than with any other metal.

Following the successful use of inert-gas metal-arc welding with aluminium, attempts were made to apply the method to other non-ferrous metals and to steels. The use of argon for welding steels was not economically attractive, but, after several years of research in the U.S.S.R., U.K. and U.S.A., techniques were developed which permitted the satisfactory use of carbon dioxide as a shielding gas. This gas is cheap and made the process competitive in many applications with established processes such as metal-arc.

The history of gas-shielded arc welding, from the first use of the helium tungsten-arc in the 1940s to the successful use of the carbon dioxide-shielded metal-arc in the 1960s has been discussed in reasonable detail because this is possibly the best introduction to this important series of processes. The impetus behind each new development can be seen in perspective and it will have been noted that the circumstances have been extraordinarily favourable for rapid exploitation.

INERT-GAS TUNGSTEN-ARC WELDING

Although electrodes of refractory metals other than tungsten have been used for inert-gas welding they are unsuitable because they erode too easily. Even tungsten electrodes are eroded, but the rate with careful usage is so slow that the electrode is justifiably considered non-consumable. The gas which surrounds the arc and weld pool must also protect the electrode.

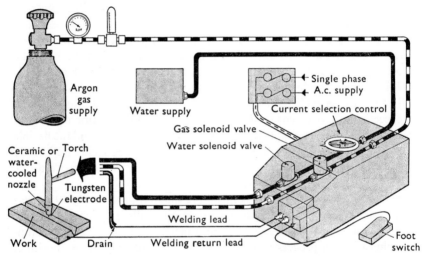

Fig. 26. Essential features of the inert-gas tungsten-arc process.

At the high temperatures reached at the root of the arc the tungsten is readily oxidized so that the shielding gas can only comprise mixtures of the inert gases and hydrogen, or in extreme cases nitrogen. Hydrogen is not generally useful for shielding because it raises the arc voltage and requires a high open-circuit voltage and it can also be absorbed by some metals giving rise to either cracking or porosity. For practical purposes, therefore, gas-shielded non-consumable electrode welding employs argon or helium for shielding and tungsten for the electrode (fig. 26).

Electrode polarity

Because of the greater heat liberated at the anode a tungsten electrode used on this polarity becomes more readily overheated than if it is the negative pole of the arc. The maximum current which the electrode will carry is reached when the molten end becomes so large as to be unstable and particles of tungsten begin to leave the electrode. Even with tungsten electrodes of $\frac{1}{4}$ in. diameter no more than 100 A can be used with d.c. when the electrode is positive. When the electrode is negative, however, the permissible current is up to eight times greater. For this reason tungsten-arc welding with the electrode positive is seldom used.

The chief advantage of the d.c. electrode positive (DCEP) method is the cleaning action exerted by the arc on the work. It is by no means certain that the widely held theory that this is because of ion bombardment is entirely correct. High-speed films of the arc indicate wide and exceedingly rapid motion of the cathode spots which have a preference for particles of oxide and other impurities. Vaporization of both oxide and underlying metal could occur at these spots, any oxide remaining being broken up and freed to float away to the edges of the weld pool. This activity of the cathode spots can sometimes be observed, particularly with aluminium, on the edges of the plate adjacent to the weld pool. Except where the cleaning action is essential, when welding aluminium or alloys containing appreciable amounts of elements forming refractory oxides, the DCEN polarity is normally used.

The characteristic motion of cathode spots is often said to cause 'in-stability' when the electrode is negative because the root of the arc can wander over the end of the electrode. Two measures are adopted to prevent this; the electrode is ground to a taper and the tungsten is doped with materials to improve its emissivity. The doping of electrodes with 1–2 per cent thoria, or with zirconia, increases the area of the cathode spot and also gives easier arc starting and improved resistance to contamination. Contamination and loss of tungsten occur largely at the start of a weld but by improving the emissivity hot spots are avoided and the electrode is made to achieve its operating temperature more easily. An additional advantage is that the current-carrying capacity of the electrode is raised.

Where the material being welded demands the electrical cleaning action of the electrode positive polarity, but currents over 100 A must be used, an a.c. power supply is employed. The a.c. arc combines the advantages of arc cleaning of the work on the half cycle in which the electrode is positive with the lower heat input to and, therefore, cooler running of the electrode when it is negative (see fig. 27). When used on a.c. the end of the electrode is not tapered and should assume a stable hemispherical shape because of superficial melting. If the current is excessive for the size of electrode this molten tip will oscillate because of the pulsating arc forces, and particles of tungsten may be ejected from a small pip which forms in the centre. Too low a welding current may not provide sufficient energy to melt the electrode end so that the arc root wanders making the arc un-stable. The optimum current range for each size of electrode depends on several factors—the design of the torch, which can influence the cooling of the electrode, the type of electrode and, possibly most important, the balance between the positive and negative parts of the current cycle (the

5

convention for reference to polarity is with respect to the electrode). As will be seen later there is a tendency for the negative current half cycle to be larger than the positive giving a d.c. component. If this is eliminated to protect the transformer from saturation the positive half cycle is increased in duration so that the electrode runs hotter and its current-carrying capacity is reduced.

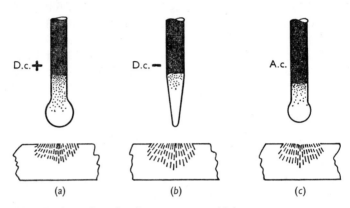

Fig. 27. Electrode and melt run contours with inert-gas tungsten-arc.

Arc maintenance

With a.c. the reversals of voltage and current introduce the problem of arc re-ignition as the arc is extinguished twice in every cycle. When the electrode becomes negative the arc ignites satisfactorily but when the voltage is reversed so that the electrode becomes positive the arc goes out and will not re-ignite unless at that instant there is sufficient voltage available at the arc gap. This problem is similar to that met in a.c. metal-arc with the difference that the voltage required is much higher for the negative/positive change with the tungsten/aluminium arc. There is also a greater contrast between the ease of the electrode positive/negative change and the subsequent reversal. This is indicated by the marked re-ignition peaks which are observed on the voltage records of the tungsten/aluminium arc shown in fig. 28.

The re-ignition can be accomplished satisfactorily in three ways. With a well-designed transformer with low electrical inertia the voltage required for re-ignition can often be supplied by the transformer giving the process of self re-ignition. If a high-frequency spark unit is used for striking the arc this can then be switched off by a relay once welding begins. Alternatively, the high-frequency spark unit may be operated from the open circuit voltage so that it ceases work when the voltage drops to that of the arc. One of the advantages of switching off the high-frequency spark is that the radio interference caused by the sparks is limited in duration.

Self re-ignition for all its simplicity has disadvantages. The open circuit voltage required tends to be high, usually approaching 100 V and the power factor has to be low because a high voltage must be available at current zero. Greater reliability is claimed for the methods in which means are specially provided for assisting re-ignition.

The voltage and current wave-forms illustrated in fig. 28 *a, b* show the cycle of events in arc re-ignition. If the voltage across the gap climbs rapidly towards the open circuit voltage the re-ignition voltage for the arc is quickly attained and the arc is restarted.

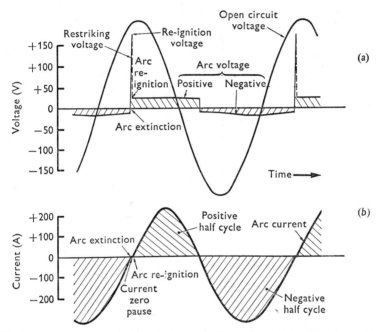

Fig. 28. Basic wave-forms of voltage and current for self re-ignition (for clarity the duration of the arc re-ignition stage and current zero pause have been exaggerated). (*a*) Voltage record; (*b*) current record.

A delay in re-striking leads to a discontinuity in the current wave-form called a current zero pause. In extreme cases, where for example there is a reduction in open circuit voltage and therefore the available re-striking voltage, the arc may not re-ignite at all on the positive half cycle. This is complete rectification, a condition quite unsuitable for welding as there is no cleaning action and the transformer can be overloaded by saturation with the d.c. The delayed re-ignition is known as partial rectification.

When left on continuously the high-frequency spark unit used for spark starting can become an arc maintainer. Re-ignition is accomplished by the sparks which discharge across the arc gap providing an ionized path for the main power circuit. Slightly lower open circuit voltages are required with high-frequency re-ignition. The spark unit comprises a capacitor, charged by a high voltage transformer, which discharges through a spark gap (fig. 29 *a*). A train of sparks is set up which lasts as long as the transformer voltage exceeds the breakdown voltage of the spark gap. This generally occurs for two-thirds of each half cycle of the mains supply. The unit, therefore, emits bursts of sparks which must be arranged to occur during

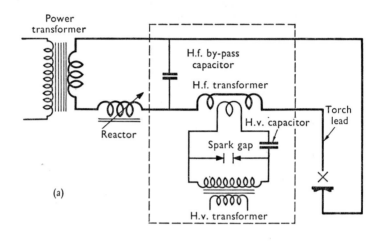

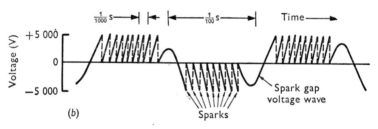

Fig. 29. High-frequency arc maintenance unit and welding circuit with basic wave-form of spark gap voltage.

the period in which the welding circuit passes through current zero. Because of the cyclic nature of the sparks re-ignition cannot be instantaneous and there is some partial rectification.

In the third method of arc maintenance a voltage surge is injected into the power circuit to supply the re-ignition peak. This is done by discharging a capacitor through a switch which is tripped automatically by the power circuit (fig. 30). When the arc is extinguished at the end of the negative half cycle the re-ignition peak begins to develop and itself fires a gas discharge valve which discharges the capacitor. Re-ignition is therefore instantaneous eliminating partial rectification, and because the transformer does not have to supply the full re-ignition voltage it is possible to reduce the transformer open circuit voltage. Welding can actually be accomplished at less than 50 V R.M.S. which permits an improvement in power factor as well as giving greater safety. The timed surge is an arc maintainer only and

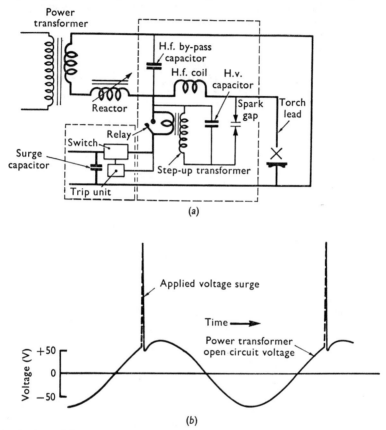

(a)

(b)

Fig. 30. (a) Surge-injector arc-maintenance unit and welding circuit; surge unit in left-hand block, spark starter in right-hand block. (b) Basic wave-form of maintenance voltage.

will not strike an arc from the cold or always after momentary extinction. A spark generator operated by the surge is used, therefore, to start the arc and is switched out automatically once the arc is started.

Direct current component

Mention should now be made of the inequality of the half cycle loops in the current wave-form. The positive half cycle tends to be the smaller because the arc voltage when the electrode is positive is higher than when it is negative. This is because the sum of the cathode and anode drops is different when the cathode is on the work from when it is on the electrode. The effect, which is known as inherent rectification, is independent of re-ignition phenomena and is observed even when the arc is running correctly. It results in a d.c. component of current which tends to saturate transformers making it necessary to derate them to about 70 per cent of their usual value. At high-welding currents it can also cause serious arc blow. Any current zero pause also causes current unbalance and adds to that caused by inherent rectification.

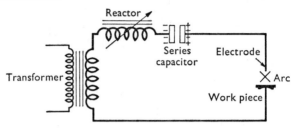

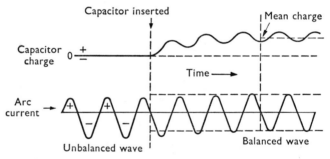

Fig. 31. The series capacitor: (*a*) circuit; (*b*) effect on unbalanced current wave-form.

These effects are usually considered detrimental so that inherent rectification is corrected by the insertion of a large capacitor in the power circuit. Banks of reversible electrolytic capacitors giving about 100 μF per amp are used. A charge is left on these capacitors when the electrode is negative which is available when it becomes positive to increase the voltage available and therefore to drive more current through the arc. The d.c. component can also be suppressed by placing large storage batteries in series with the arc so that the d.c. component is biased out (fig. 31).

Inherent rectification or the d.c. component has an important thermal effect in welding. Because it is the result of lack of balance between positive and negative half cycles it affects the distribution of heat between work and electrode. The electrode runs hotter when the inherent rectification is balanced out and the current on the positive half cycles is increased. Conversely, more heat is available in the work when there is no d.c. suppression. With weld beads made at the same current setting and speed, it will be noticed that those where d.c. suppression is used can be 15 per cent narrower. In the U.S.A. where d.c. suppression is less common than in Europe the recommended welding conditions tend to higher currents for a given size of electrode and faster welding for the same current.

Starting the welding arc

Contamination and tungsten loss can be greatly reduced by striking the arc with a low current through a pilot circuit and switching to the main current a few cycles later when the electrode has warmed up. In the absence of such techniques, and whether touch starting is used or not, it is helpful to strike the arc on a piece of scrap material and then re-strike on the work once the electrode has reached operating temperature. These measures are particularly important when welding at high currents because the chances of losing tungsten to form tungsten inclusions are greatly increased during the warming-up period of the electrode.

During the warming-up of the electrode on a.c. and until it reaches a temperature which will permit thermionic emission it is possible for the arc to fail on the electrode negative half cycles. This leads to rectification in the opposite sense to that normally encountered while the arc is running so that the occurrence is known as 'inverse rectification'. Once the electrode is hot the arc re-ignites readily with its cathode on the tungsten and it is more likely to extinguish when the work is negative. The most important effect of inverse rectification is that it leaves a charge on the d.c. suppression capacitor of the reverse polarity to normal. This opposes the half cycles of electrode negative current, which are required to heat the electrode, so that the arc puts itself out. To prevent this, the suppressing capacitors are often switched out when the arc is being struck and switched in later when it is running normally.

Welding techniques

After the arc has been struck the torch is held stationary while the molten weld pool is formed. If the welding current is adequate this should not take more than a few seconds and the surface of the pool should be bright and clean. A leftward technique is used with the torch held at 80° to give visibility of the pool and to sweep argon ahead of the bead. Once the pool is established welding can proceed and filler metal may be added if required. The action of introducing filler metal to the weld pool can disturb the gas shield and entrain air. It is helpful, therefore, to keep the end of the filler rod within the shield at all times. This also prevents oxide forming on the end of the filler rod.

If the filler rod diameter is too small in manual welding it will melt rapidly forming a globule at its end. Conversely, too large a rod may disturb the arc and cause oxide inclusions by shielding the weld from the arc-cleaning action. Filler rod of the correct size can be held close to the pool and no violent motions are necessary when adding metal. With mechanized welding filler wire from a coil is fed into the leading edge of the weld pool and is arranged to make good contact with the solid metal just ahead of this edge. Smaller diameter wires can be used than for manual welding where this contact cannot be maintained.

Mechanized welding is widely used with the tungsten-arc and frequently to simplify the process the joints are designed so that the addition of filler metal is unnecessary. This requires the use of edge joints or accurately machined and fitted butt joints. When the latter are used it is usual to allow the weld bead to sink slightly into a backing bar or under gravity so that oxide films at the root are broken. Even so, it is sometimes observed that the weld occupies the full cross-section of the joint, although no extra metal has been added. This paradox is noticed when heat spreads ahead of the pool so that metal in this region expands and upsets. The extra metal is gathered into the pool, increasing the thickness in the weld bead and leaving transverse shrinkage or deformation in the welded component. Although the addition of filler metal is avoided if possible it is sometimes necessary so that the composition of the weld metal can be changed to avoid cracking or porosity.

Stopping the weld

When the current is cut off and the arc extinguished the electrode and weld pool begin to cool, but an adequate flow of argon must be maintained until the danger of oxidation is past. In most tungsten-arc welding equipment the flow of gas is controlled by a solenoid valve which is timed to open before the arc ignites and to close after allowing sufficient time for the electrode to cool.

When the arc is extinguished abruptly at full welding current the weld pool solidifies with a central pipe, a defect which can cause leaks in joints intended for service in vacuum or under pressure. The crater pipe is overcome by reducing the current gradually before switching off with a device known as a crater filler. In this way the pipe is fed with liquid metal and the crater solidifies progressively. It is possible to reduce the crater in mechanized welding by increasing the speed before switching off. With manual welding the welder should feed the crater with filler metal. In d.c. welding the arc may be extinguished by lengthening the arc gap, causing the voltage to rise and the current to drop prior to extinction in a manner depending on the volt–amp characteristics of the power source.

Applications

The tungsten-arc process is used with welding currents from $\frac{1}{2}$ A up to 700 or 800 A and is one of the most versatile methods of welding. The lowest currents are used with delicate air-cooled torches to weld metal

0·002 in. thick. For these very low currents DCEN is usually used. A.c. is employed between 25 and 350 A, but above this current the risk of tungsten inclusions increases. Up to 800 A DCEN is again used and these high currents are employed particularly for welding thick copper. Torches designed for up to about 100 A are entirely air cooled but for higher currents water-cooled torches must be used.

Although high welding currents permitting the welding of thick metal are possible, tungsten-arc welding is primarily a process for welding sheet metal or small parts. The process is at its best when welding single pass or double-sided close butt joints, edge joints or outside corner joints. It is less suitable for fillet welds with which care must be taken to obtain good fusion into the root. Because it is so easily mechanized and gives high-quality welds the process is greatly favoured for precision welding in the aircraft, atomic energy and instrument industries. Circumferential and edge welds, for example can-sealing joints, are very suitable for mechanized tungsten-arc welding. Arc-length control systems are sometimes used with mechanized welding in which the arc voltage provides a signal to raise or lower the welding head so that the arc can follow a curved or undulating surface.

Various automatic devices are available for welding tubes to tube-plates. Typical of these is the method in which a miniature torch revolves around a central spigot which is fitted into the tube.

Special forms of tungsten-arc

Pulsed tungsten-arc. For welding thin sheet material advantage can be taken of a d.c. power supply in which the current is modulated. A low current is used to maintain the arc and a square wave pulse superimposed of sufficient magnitude to melt a weld pool. If the torch is moved uniformly along the joint a series of nuggets is fused in the work. These may be separated or overlapped to produce a continuous seam. The advantages claimed for this technique are that it is easier to control penetration when manually welding thin material and heat dissipates between pulses so that a build-up of heat is avoided.

Twin tungsten-arc. Where a smooth wide weld bead is required, for example in the manufacture of cable sheathing from strip, twin electrodes have been used. These are connected to a Scott two-phase transformer so that the arc strikes from each to the work alternately. When electrodes arranged in this way are moved side-by-side along a joint a smooth weld without undercut is obtained.

Tungsten-arc spot welding. An important variation of tungsten-arc welding is used for arc spot welding. The equipment is basically the same as that for conventional manual tungsten-arc welding except that the control system includes a timing device and the torch is modified, in particular the gas nozzle. The process is used to join overlapping sheets of equal or unequal thickness. With both sheets firmly in contact and the torch nozzle pressed against the upper sheet a DCEN arc is struck for a predetermined time. A nugget is melted in the upper sheet which penetrates to the lower sheet fusing the two together. It is important, of course, that the oxide on the faying surface is fused or dispersed otherwise a weak joint will result. Only clean surfaces should be used.

The process is not generally used for aluminium alloys because the oxide film is not easily dispersed. Increasing the turbulence in the pool with a stronger plasma jet, however, by using high currents for shorter times allows aluminium sheets to be welded. With this technique the top surface of the nugget tends to have a deep hole which for some purposes is regarded as unacceptable.

Generally the top surface of tungsten-arc spot welds is free from excessive depressions as a crater-filling device is used. There is even frequent evidence of the gathering effect, mentioned in connection with close butt welds, which leaves the centre of the nugget slightly raised. Where the crater must be filled and the nugget raised appreciably or when it is necessary to control weld pool metallurgy, a separate addition of filler wire may be made.

Tungsten-arc spot welding is useful where access can be gained to only one side of a lap joint or where the bulky head and high pressures associated with resistance spot welding cannot be used. See also gas metal-arc spot welding.

PLASMA-ARC WELDING

The plasma arc heat source may be considered as a development of the inert-gas tungsten arc. There are two types, the non-transferred and the transferred or constricted arc. If the inert-gas tungsten-arc torch is provided with a separately insulated water-cooled nozzle that forms a chamber round the electrode and the arc is struck from the electrode to this chamber, the arc plasma can be expelled from the nozzle in the form of a flame. This is the non-transferred arc which, with a powder feed into the plasma, is used for metal spraying.

When the arc is struck instead from the electrode to the workpiece the arc is contracted as it passes through the orifice in the nozzle—this is the transferred plasma arc and it is used for cutting because of the high energy density and velocity of the plasma. To start an arc of this kind it is necessary to have a pilot arc from the electrode to the nozzle.

The transferred arc can be used for welding as well as cutting and is used with two different techniques. At low currents for welding sheet metal less than $\frac{1}{16}$ in. thick and at currents up to 400 A for welding thick metal using the deep penetrating key-hole technique. When using the plasma arc for welding additional inert-gas shielding is required which is provided by an annular flow concentric with the primary gas flow through the plasma-arc chamber. A d.c. electrode negative arc is generally used except for aluminium where the polarity may be reversed. Arc voltages are higher than for tungsten-arc welding and may rise to 50. The orifice through which the plasma passes is about 0·10 in. diameter.

The deep penetration technique for using the plasma arc on thick material has certain similarities with the electron-beam process. Because of the high energy density of the plasma arc a hole is melted through the metal being welded and as the torch is moved along the joint energy is transferred from the plasma to the walls of the hole which is carried along the joint. It is from this feature that the name 'key-hole technique' is derived. A cross-section of such a weld shows a wine glass shaped or fingered penetration, the 'stem' being a result of heat transfer to the walls of the hole and the 'bowl' shape on the surface because of surface heating in the normal way associated with arc welds. As with the electron-beam process the relative proportions of 'bowl' and 'stem' depend on the uniformity of energy in the beam. The preferred shape of penetration giving the minimum of 'bowl' can be achieved by ensuring that the plasma is as near as possible a parallel stream without a low velocity, cool fringe. Arcs tend to spread near the workpiece surface and it is helpful therefore to focus the plasma. This is done by another gas flow in between the plasma flow and the annular shielding flow. Several jets of gas are directed at an angle on to the plasma arc a short distance after it leaves the nozzle so that any tendency to divergent flow is corrected.

GAS METAL-ARC WELDING

In gas-shielded arc welding with an electrode having a lower melting point than tungsten the electrode end melts and molten particles are detached and transported across the arc to the work by the magneto-dynamic forces and gaseous streams. Unlike tungsten-arc welding gas-shielded metal-arc welding is invariably used with d.c. and usually with the electrode positive. It will be recalled that DCEP welding permits the arc cleaning action to take place but that this polarity cannot be used with a non-consumable electrode because it becomes overheated.

The size and frequency of drop transfer across the arc is related to wire diameter and current. Voltage has only a limited effect except with the short circuit process (see p. 85). Drop size decreases and transfer frequency increases with an associated improvement in directness of transfer as the current density is increased. In gas-shielded welding, as there is no flux covering to provide a shroud at the electrode tip to direct the transferring metal, advantage must be taken of the favourable characteristics of small diameter electrodes with high current densities. The size range employed is usually between 0·030 and 0·080 in. diameter. Apart from metal transfer characteristics small diameter wires are required for improved arc-length control and ease of handling.

Arc-length control

With an automatic welding head arc-length control can be by the controlled arc system. This has been, and is still, used for gas-shielded welding but the self-adjusting arc method in which wire electrodes are fed at constant speed has the great merit of simplicity. A constant rate of wire feed is also preferable where the electrode wire must be pushed down the flexible conduit of semi-automatic equipment. In fact, it is the self-adjusting arc system with its constant wire feed that has made possible the wide use of semi-automatic equipment in which the consumables, such as gas and wire, are fed through a flexible conduit to a hand-held torch (fig. 32).

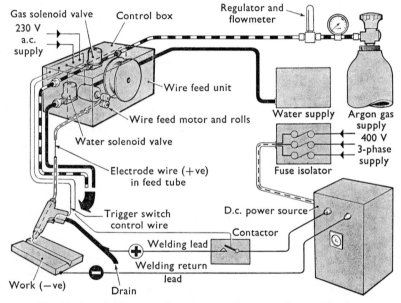

Fig. 32. Essential features of semi-automatic gas metal-arc equipment.

How can arc-length variations be corrected without a change in feed rate? The answer lies in what happens to the burn-off rate of the wire. For a stable arc length the burn-off rate must equal the wire-feed rate. If the burn-off rate is decreased the end of the electrode will advance toward the work at a speed equal to the difference between the old and new burn-off rates. The converse occurs if the burn-off rate is increased. Burn-off rate is determined by welding current (according to the burn-off characteristic, for example as shown in fig. 33) and current is determined by arc voltage (according to the volt–amp characteristic of the power source). Voltage in its turn depends on the arc gap.

Examine what happens when an arc of length l_1 is disturbed by having the welding torch suddenly drawn away from the work. Momentarily the arc length is increased to l_2 which, as fig. 14 (p. 34) shows, increases the voltage and reduces the current according to where the new arc characteristic meets the power source output curve. Referring now to fig. 33 it will be seen that a decrease in current causes a decrease in burn-off rate. This decrease in burn-off rate upsets the balance between the burn-off rate and the feed rate so that the latter becomes the greater of the two. The end of the electrode, therefore, advances toward the work and the equilibrium arc length is restored. These events are completely reversible.

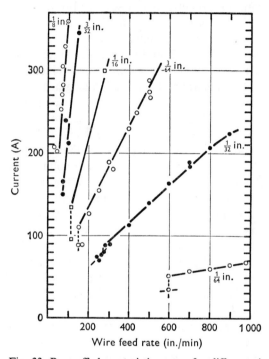

Fig. 33. Burn-off characteristic curves for different sizes of aluminium wire in argon.

It will be clear from fig. 33 that the change in burn-off rate for a given change in current will depend on the slope of the burn-off rate characteristic. The curves drawn are for different diameters of aluminium wire of the same composition, and the $\frac{1}{32}$ in. diameter wire, for example, gives a higher burn-off rate for the same change in current than the $\frac{1}{16}$ in. diameter wire. For semi-automatic welding the rate of change in burn-off rate must be greater than the rate at which the torch would be moved because of unsteadiness in the welder's hand or of changes in the contour of the work. If this is not so the arc length will not be held steady. It is now clear why semi-automatic equipment using a self-adjusting arc employs fine wires.

As the static output characteristics have shown the power source itself has an important part to play in self-adjustment. If the characteristic curve is steeply drooping the change in current (and hence burn-off rate) for a given voltage change is less than if the curve is relatively flat. Flat characteristic or constant potential power sources are, therefore, preferred for the self-adjusting arc process as they give increased self-adjustment.

Arc characteristics

The use of a constant potential power source also permits a simplification of the operating procedure. With a drooping characteristic power source the current is determined largely by the current setting on the power source, as explained in chapter 2, and the wire-feed rate must be adjusted until an arc of the appropriate length has been obtained. Increasing the wire-feed rate shortens the arc and a reduction lengthens the arc and vice versa. When a constant potential power source is used the voltage range over which the process will operate is greatly restricted. Figure 14 shows that to be able to operate at the same current and voltage as with the drooping power source it is necessary to reduce the open circuit voltage. Open circuit voltage determines the working arc voltage, which can be 1–4 V less than the open circuit voltage because of the electrode extension beyond current pick-up, contact resistance and lead losses. Variations in wire-feed rate now cause changes in current corresponding to those indicated on the burn-off curve (fig. 33), with almost no change in voltage or arc length. It is easier, therefore, to set and maintain welding conditions with the constant potential power source as the voltage, and hence the arc length can be predetermined and the current automatically rises to that required to burn-off the wire at the appropriate rate. The wire-feed rate control, therefore, also controls current, the complete relationship between current, voltage and wire-feed rate is shown in fig. 34 in what is known as Z curves.

Effect of current. Arc-length control is only one of several aspects which must be considered in gas metal-arc welding. The way in which metal is transferred from the end of the electrode to the work is also of considerable importance. Metal leaves the electrode wire as drops; the size, frequency and directionality of the transfer being features influenced by the process variables. For any diameter of wire the frequency with which drops are detached varies with current, as shown in the transfer frequency curves in fig. 35. Each metal or alloy has a typical curve which may, however, be altered if the composition of the shielding gas is changed. Marked differences exist, for example, between the curves for steel in argon–oxygen mixtures and in carbon dioxide. Curves such as those in fig. 35 are plotted from oscillographic records of voltage and current supplemented by high-speed cinematography. As each drop leaves the wire a neck is formed between drop and wire momentarily giving a deflection in the voltage and

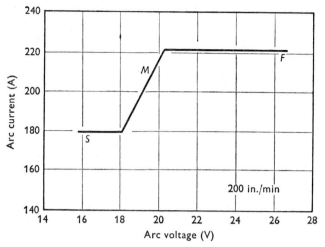

Fig. 34. 'Z' curve showing the current, feed rate and voltage relationship for 0·064 in. diameter aluminium wire in argon. *F*, Free flight zone; *S*, short circuit; *M*, mixed type transfer zones.

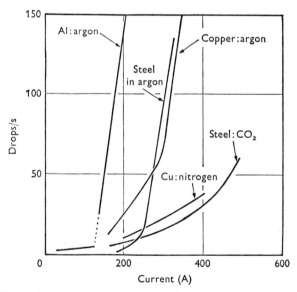

Fig. 35. Effect of current on the frequency of drop detachment with 0·064 in. diameter wire for a number of gas metal-arc systems.

current records in the form of blips. Welding conditions here were chosen so that the arc length is sufficiently long for the drops to detach without bridging to the work. This is known as the free-flight condition (see fig. 10*a*, and oscillograms in pl. 5). Transfer frequencies can vary from a few per second with thick wires or low currents up to several hundred per second at currents over 300 A. As the drop transfer frequency rises the drop size and volume are reduced in proportion. From the welding point of view frequencies in excess of 20/sec are generally required to give smoothness of operation and the formation of a continuous weld bead.

The frequency/current curve for aluminium shown in fig. 35 and the corresponding current/feed rate or burn-off curve (fig. 33) are probably unique in that there is a discontinuity at low current when both the arc and transfer type change abruptly. Above this transition point or threshold current the drops are detached and transferred by the plasma jet with considerable velocity. The electro-magnetic forces keep the arc on the axis of the electrode so that it may be directed at will irrespective of the angle made with the plate. Overhead and vertical welding is therefore feasible, although the absence of flux to hold up the metal is a disadvantage except with aluminium alloys. This is often known as a spray-type arc but it is more accurately described as a free-flight type arc. At very high welding currents—above 400 A the strength of the plasma jet is increased so much that unless special precautions are taken air may be entrained to form oxide films in the turbulent weld pool and the weld bead is said to be 'puckered'. With metals other than aluminium which do not form refractory oxides puckering may not occur but there is evidence that the arc force with welding currents over 700 A can cause metal to be blow out of the weld pool. Below the transition current large drops grow on the wire, the arc envelops only the lower half of the drop and transfer is by gravity. This condition is called 'subthreshold' or globular transfer and is unsuitable for welding.

Effect of wire diameter. The transfer frequency curves shown in fig. 35 change for different wire sizes. Wires thinner than $\frac{1}{16}$ in. have curves displaced to the left; thicker wires have curves displaced to the right. The transition currents for aluminium vary with wire diameter in a manner indicated by the relationship $I_c = 1730d + 30$, where d is the wire diameter in inches and I_c the transition current below which spray transfer ceases to exist for a 22 V arc. As the welding current should be in excess of the transition current to obtain a free-flight arc, welds in thin material requiring low currents must be made with small diameter wires. With metals other than aluminium the change from spray to globular is less well marked, but transfer characteristics generally deteriorate when the drop diameter exceeds the wire diameter.

Metal transfer characteristics. The behaviour of a gas-shielded metal-arc is not described wholly by the burn-off and transfer frequency curves. Several different types of metal transfer have been observed, each of which is characteristic of a particular metal or group of metals. From the welding point of view much depends on the way in which the drop is transferred to the pool. With an aluminium or stainless steel metal-arc in argon the transfer takes the form of a stream of drops projected axially from the end of the wire. There is a notable difference in electrode behaviour in an inert

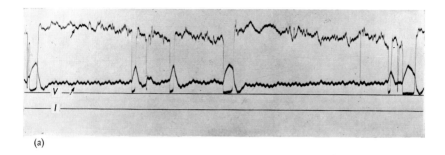

(a)

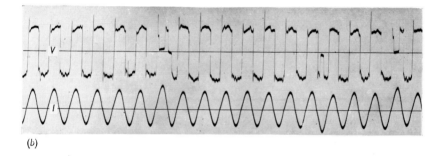

(b)

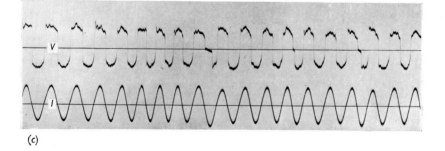

(c)

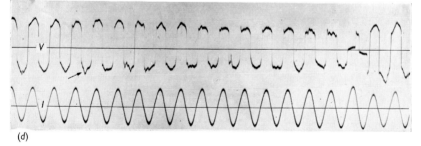

(d)

1   Voltage and current oscillograms of metal-arc welding. (*a*) Class 1, cellulosic
DCEP. (*b*) Class 1, a.c. (*c*) Class 2, rutile. (*d*) Class 6, lime-spar low hydrogen.

2 Operator making an overhead weld in a heavy structural member for a steel-framed building.

3 Nozzles welded into the dome of a large pressure vessel for a nuclear reactor.

4 Self-propelled automatic welding machine at work in a shipyard using a continuous covered electrode.

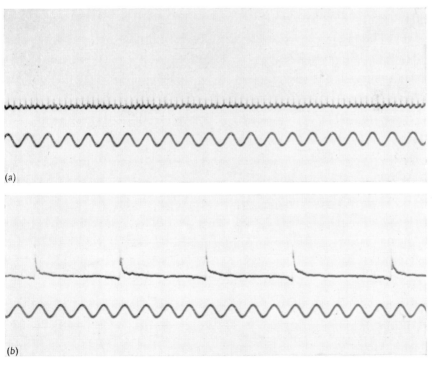

(a)

(b)

5 Oscillograms of arc voltage showing metal-transfer types: (a) spray-type transfer, 180 transfers/s, $\frac{3}{64}$ in. diameter wire at 180 amp; (b) subthreshold-type transfer, 12 transfers/s, $\frac{3}{32}$ in. diameter wire at 150 amp.

6 Photomicrograph of capacitor discharge stud weld, dark line indicates fused metal; × 10.

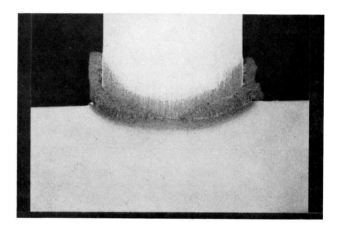

7 Photomicrograph of cross-section of arc stud weld using fig. 44*d* type stud; × 5.

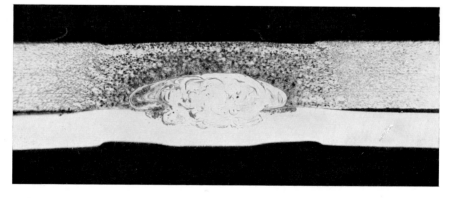

8 Nugget formed in a resistance seam weld between a steel, lower, and a nickel–iron alloy, upper. Turbulence in the molten nugget due to electro-magnetic forces is clearly visible; × 15.

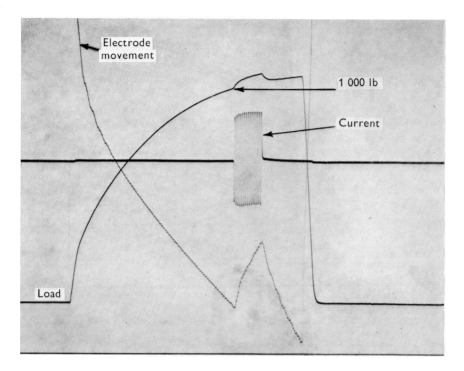

9 Records of current, load and electrode movement in making a spot weld in two thicknesses of 0·064 in. thick mild steel at 5·8 kA. Movement record indicates nugget expansion.

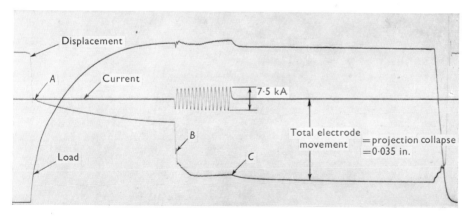

10 Records of current, load and electrode movement in projection welding two 0·064 in. thick mild steel sheets. Cold collapse begins at *A*. Note check in hot collapse at *B* due to current zero between half cycles, also gradual increase of current from 6·8–7·5 kA. Slight upward movement before *C* due to expansion and downward movement after because of weld cooling.

11 Flash-welding clamps and back-stop arrangements. Platens in the position after welding but with upper clamps released. Pipes going to clamps are for water cooling.

12 Double-operator welding of a copper vessel. Weld runs are made simultaneously on each side of the joint.

13 Iron Pillar of Delhi.

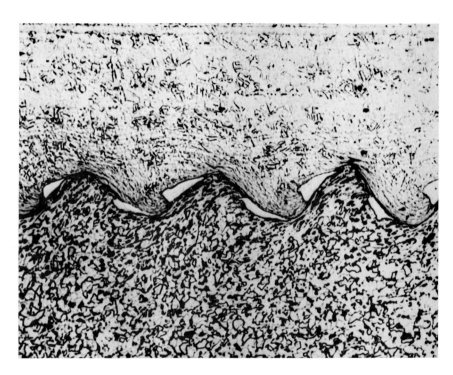

14 Photomicrograph of weld zone between stainless steel (top) clad explosively to a carbon steel base; ×150.

atmosphere between aluminium and copper base alloys and ferrous metals. With the non-ferrous metals the electrode end tends to remain spherical and the particles transferred are also spherical. Ferrous materials in an inert atmosphere, however, form a pointed electrode end from which spherical drops stream off into the arc and are transferred to the weld pool. Transfer in a nitrogen- or CO_2-shielded metal-arc is less satisfactory. With these dissociable gases the arc root does not envelop the end of the electrode so that the detachment and transport of the drop by the plasma jet in the manner proposed by Needham, Cooksey and Milner, cannot take place. Instead, the arc root is broadly based, the drop grows larger than the wire diameter and appears reluctant to leave the wire as if a force existed repelling the drop from the pool. This force might be electro-magnetic, caused by lack of symmetry in the current flow, or it may be a result of a plasma jet operating away from the pool and competing with the anode jet. As a result the transfer tends to be non-axial and naturally such systems are prone to spatter.

Spatter can be minimized by selecting the arc voltage appropriate to the current being used. At currents of less than 400 A a short low-voltage arc helps because large drops which may be deflected away from the pool cannot form as the drop bridges to the weld pool and is detached. This type of transfer is not strictly spray or free-flight. With the CO_2-shielded steel arc too short an arc at high currents, however, results in the arc becoming buried in the weld pool crater. A long pointed liquid stream develops to the electrode end which whips about because it is unstable in its own electric-field and this causes severe spatter. This condition is rectified by increasing the voltage slightly which produces a globular end to the electrode. The relationship between current and voltage for optimum smoothness of transfer and minimum spatter is indicated in fig. 36. The deflected or non-axial type of metal transfer is also a characteristic of electrode negative operation because the arc tends to form a small cathode spot on the electrode. Coatings of the alkali metals on the electrode surface improve the emissivity and can cause the arc to envelop the electrode end allowing the satisfactory plasma jet metal transfer to take place.

The discussion of spray transfer welding will have indicated that the method is usable over a comparatively narrow current range—the upper limit set by the effects of arc force, the lower limit, by the onset of globular transfer when large drops are detached non-axially or fall by gravity and positional welding is impossible. This lower limit is influenced by the diameter of the electrode wire so that there has been a tendency to reduce wire sizes in an attempt to make the process work with lower currents and therefore to be applicable to the welding of thinner material. Two difficulties are encountered in such attempts: first, fine wires are difficult to feed through the welding equipment, particularly flexible conduits, and secondly, even though the current is reduced for smaller wires, the wire-feed rate required may well be excessive. In addition, as fig. 33 shows, the smaller the wire diameter the narrower the range of current over which that size is usable.

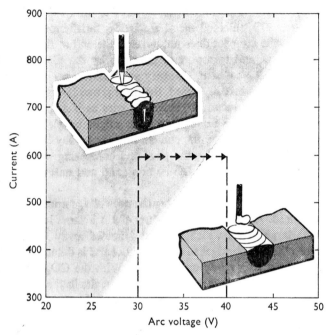

Fig. 36. Influence of arc voltage on metal transfer characteristics with high current CO_2 welding of steel.

Features of gas metal-arc equipment

Gas metal-arc welding equipment comprises power source, wire-feed unit, current contact and gas barrel, and a control system for switching on and off welding current, gas, wire and cooling water. The general characteristics of power sources and the use of self-adjusting and controlled arc systems have been discussed in previous chapters. With automatic welding equipment the wire feed unit and the current contact and gas barrel are combined in a single welding head. For semi-automatic welding flexibility is generally achieved by separating the wire feed unit from the torch and passing wire, gas, current and cooling water through a flexible conduit. To be able to push wire for several yards down a flexible tube it is necessary to have high-powered wire-feed motors driving non-slip knurled or 'V' groove rollers. The wire-feed tube must have a smooth bore, fit the wire sufficiently closely to prevent buckling of the wire and, although flexible, must be rigid enough not to kink when bent to a tight radius. Ferrous metal wires can be fed through the spiral steel wire-feed tubes readily, but aluminium and some non-ferrous metals are abraded and jam unless the feed tube is lined with extruded nylon or P.T.F.E. The wire itself must be fed from reels without the risk of snagging and careful layer winding is normally adopted. Wire which is drawn to an excessively hard temper or which has a twist is difficult to handle because the reels tend to be springy and the wire spirals on leaving the contact tube, causing a wavy weld bead. Reels are friction loaded so that wire is unwound under slight tension. To prevent an over-run on wire feed because of inertia the wire-feed motors may be locked mechanically when the current is switched off.

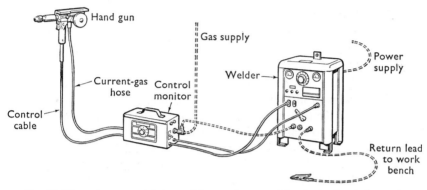

Fig. 37. Fine-wire gas metal-arc welding equipment with wire-feed unit in torch.

Wire feeding difficulties increase with a reduction in the diameter of the wire so that there is a limit to the policy of lowering the usable current by using smaller diameter wires. The use of wires as small as 0·035 in. diameter is feasible with ferrous metals but with aluminium alloys the wire-feed difficulties become serious below 0·062 in. diameter. With aluminium alloys fine-wire welding is feasible only if the length of the feed tube is severely reduced or the need for a tube is eliminated. Figure 37 shows fine-wire welding equipment for aluminium in which wire-feed motor and a small reel of wire are mounted in a hand-held torch. The wire-feed motor is inside the torch handle.

In almost every gas metal-arc equipment the welding current is introduced to the wire by passing it down a copper tube. Early equipment included pressure devices to give good contact, but it is now usual with the semi-automatic equipment to rely for contact on the fact that because it has been coiled the wire is never straight and contact is made in at least three places within the contact tube. A variation in the point of current pick-up can occur, however, which alters the resistance between contact and arc and may cause variations in burn-off rate because of its effect on overall circuit resistance. When welding with high currents or with high-resistance metals the current contact tube may be shortened or fitted with a smaller diameter tip to reduce this variation. The detachable tip has the advantage of being replaceable in event of damage by the arc.

Equipment for use above about 250 A employs water to cool the gas nozzle. Water cooling and a chromium-plated surface make the removal of fume and spatter from the nozzle easier. Heavy duty automatic welding heads also have water-cooled contact tubes.

Among the control devices frequently fitted to gas metal-arc equipment are solenoid valves to turn gas and water on and off, and relays to switch in and out wire-feed motors and current contactors, according to the voltage between wire and work or in response to a switch.

Applications

With welding currents of less than 250 A gas metal-arc welding is used for all-position welding with semi-automatic equipment. The process is indispensable for non-ferrous metals with which flux-shielded processes are generally less satisfactory because of the complications of weld pool chemistry and poor metal transfer characteristics. Using fine-wire or pulse techniques for the thin material, non-ferrous metals can be welded from sheet gauges up to massive thicknesses with suitable choice of edge preparation. Without the special techniques the lower thickness limit is about 0·15 in. For steels the competitive process to gas metal-arc at currents below 250 A is manual metal-arc.

Higher welding currents than 250 A are used only with downhand welding techniques and if the joints are straight or are simple circumferential seams automatic equipment may be used. Otherwise, up to 450 A semi-automatic manual equipment would be used and, as with low-current all-position welding, the process in this range is indispensable for non-ferrous metals. On steels it is particularly useful in bridging the gap between manual metal-arc and the automatic methods. The maximum current for automatic gas metal-arc is about 700 A.

The main features of the gas metal-arc process are: suitability for non-ferrous metals, ease of positional welding, use as a semi-automatic process, absence of fluxes, cleanliness and ease of mechanization. As the discussion earlier in the chapter will have indicated, gas metal-arc welding is possibly the most widely used process in terms of range of metals and applications if not in amount of welding. The industries served include shipbuilding, general and heavy electrical engineering, the aircraft engine, pressure vessel, tank, pipe, domestic equipment and automobile manufacturing industries.

Gas metal-arc spot welding

When fitted with a timer, and a special gas nozzle gas metal-arc equipment can be used for arc spot welding overlapping sheets or tacking butt or lap joints as shown in fig. 38. A vented metal nozzle of a shape to suit the application is fitted to the contact and gas barrel so that the torch can be held against the work. The process is operated for periods of 1–5 s and melts a slug between the parts to be joined. A critical part of the process is arc initiation which must be reliable and consistent. This is aided by a flat characteristic power source and clean surface to the work. The metal transferring from the wire scours deeply into the crater, breaking up oxide films at the faying surfaces so that the process can be used on aluminium as well as steels. The use of a consumable electrode has two other advantages for arc spot welding as well as that just mentioned. Because metal is added to the pool the top surface of the spot weld is raised and should be free from piping. Where it is necessary to adjust weld pool

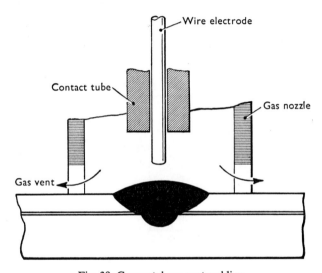

Fig. 38. Gas metal-arc spot welding.

composition to control cracking, porosity or strength a wire of different composition from the work may be used.

Overlapping sheets up to 0·125 in. thickness can be welded without preparation but from 0·25 in. upward it may be necessary to drill a hole in the top sheet to ensure penetration and prevent excessive build-up of metal. This technique can also be used for welding dissimilar metals, for example plug-welding steel to non-ferrous metals.

Short-circuit and pulse-transfer techniques

For those metal-gas systems in which the arc root envelops the end of the electrode there is a possibility that a pulsed-arc technique may permit an extension of gas metal-arc welding to lower current ranges. This method operates by inducing spray or free-flight transfer at average currents where the arc forces would not normally be sufficient to detach and propel drops across the gap. Two levels of current are used, a relatively low current to melt the wire and a pulse of greater magnitude to detach the drop. The burn-off rate approximates to that for the average current but the metal transfer characteristics are those of the pulse current. Pulse duration must be long enough for the processes of envelopment and plasma jet action to develop. Apart from the extension of gas metal-arc welding to working currents below threshold the method may help to improve the metal transfer characteristics of gas/metal-arc systems in which transfer is not completely satisfactory.

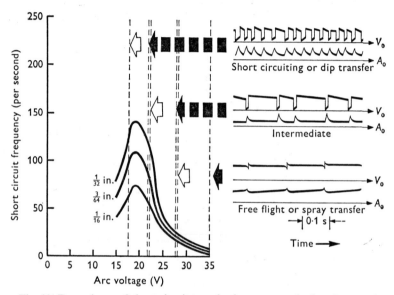

Fig. 39. Dependence of short-circuit transfer frequency on both voltage and wire diameter for CO_2 welding of steel.

Another process exists, however, in which the low-current difficulty is overcome by employing an entirely different welding technique in which metal is transferred by obliging each growing drop to touch the weld pool. The technique, known as short-circuit or dip-transfer welding, has been used particularly for the welding of steel in a carbon dioxide atmosphere but is also applicable to other gas/metal systems.

If the arc voltage is reduced progressively while keeping the wire-feed rate constant a change can be recorded in the rate at which drops are detached from the wire. This relationship between voltage and drop transfer frequency is shown for the steel–carbon dioxide system in the F/V curve (fig. 39).

Reference has been made to arc voltage in connection with the F/V curve but this is not strictly accurate as the transfer mechanism changes from free flight at the extreme right of fig. 39 to short circuiting at the left.

As the latter is an intermittent arc it is more convenient to plot transfer frequency against the voltage applied between wire and work, which is in effect the open circuit voltage. For reasons explained later a constant potential power source must be used for the short-circuiting arc so that the arc voltage is close to the open circuit voltage. A reduction in voltage requires a shortening of the arc gap so that drops are detached before they reach the size normally associated with the current drawn. The short cir-

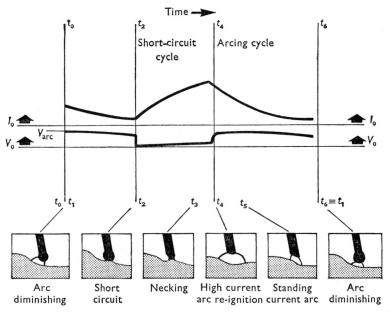

Fig. 40. The short-circuit or dip transfer process. Upper: typical voltage and current records; below: appearance of process from high-speed cine record. (Steel wire $\frac{3}{64}$ in. diameter fed at 100 in./min in CO_2 atmosphere.)

cuiting becomes a well-marked cyclic process as the voltage is reduced further, eventually reaching a maximum frequency. At still lower voltages the process becomes unstable as an arc is difficult to maintain for even a short time and the wire stubs into the work.

A complete short-circuit cycle illustrated in fig. 40 is as follows. With an arc established the drop begins to grow. This occurs with a falling current so that the wire-feed rate exceeds the instantaneous burn-off rate and the arc gap closes. Eventually the drop touches the pool which rises slightly when the molten bridge of metal is formed. The arc is extinguished and because of the reduced resistance between current contact and work now existing the current begins to rise and the voltage drops to a low figure. Resistance heating takes place in the wire and bridge and finally the current is sufficient for the electro-magnetic pinch effect to rupture the bridge.

Other forces such as electrode motion and surface tension also play a part. With a gap formed the voltage rises as an arc is established and the current decays to the minimum or standing current value. As the arc is formed a plasma jet develops which helps to drive the separated metal and weld pool downward. The force of the plasma jet decays with the current and the whole cycle is then repeated. The shape of the oscillograms of current and voltage rise and decay is determined by the reactive inductance of the circuit, energy being stored during the period of current rise and released in the arc. It is also necessary that there should be an adequate rise in current on short circuiting otherwise the bridge formed will freeze and not be broken. This condition can only be obtained with a flat or near flat characteristic to the power source.

A crucial stage of the cycle is the point at which the arc and plasma jet are established. If the current rises rapidly and reaches too high a value the arcing stage is explosive and molten metal in the weld pool is splashed out as spatter. Should the current rise be too low there is failure to fuse and separate the molten material before the cold electrode wire is plunged into the pool so that stubbing occurs. The current surge depends on the circuit characteristics, both resistance and reactance being important. In practice the desired characteristics are achieved by using motor-generator or transformer-rectifier welding sets having a variable slope or a constant potential output in conjunction with a variable or tapped reactor. Because fine adjustments are often required the infinitely variable reactor is usually preferred. Resistance is affected by electrode wire diameter, electrode extension beyond the contact tube and by the wire composition. All three variables influence short-circuit behaviour.

The greater the wire diameter the lower is the peak on the F/V curve and the larger is the inductance which must be added to the circuit to control the spatter produced in the short-circuiting process (fig. 39). Although the short-circuit rate is lower for the larger diameter wires there are often good reasons for its use. In the CO_2 welding of steel, for example, a wire of 0·030 in. diameter gives a high short-circuit rate. Because the arc is physically small, however, and the arcing periods during which penetration is achieved are short this size of wire is limited to the welding of thin sheet. When the dip transfer technique must be used on heavier plate, for example when positional welding is necessary, there is an advantage in using wire of $\frac{1}{16}$ in. diameter. With this wire the arc is larger and greater penetration is achieved because the arcing part of the cycle is relatively larger and the standing current higher. It has been shown that for the steel/carbon dioxide system, there is an optimum set of operating conditions for each diameter of electrode wire. The variables which must be controlled are: wire-feed rate, open circuit voltage and the circuit characteristics. It is suggested that the latter can be summarized by the rate of current rise after a short circuit, a property termed the response rate.

Electrode extension is particularly important with the smaller range of wire diameters. An increase in electrode extension, which results if the welding head is withdrawn from the work, causes an increase in resistance heating of the wire. The melting rate is increased and large globules form on the wire, the short-circuiting rate is reduced and the process becomes unstable.

A change in material composition has a marked effect on the short-circuiting process. Probably arc type, melting point and thermal and electrical conductivity are all important factors in determining how well the short-circuiting process operates with a metal. Short-circuit welding works particularly well with mild, alloy and stainless steels but can be shown to work with nickel, magnesium, aluminium bronze and alluminium alloys. With the higher thermal diffusivity metals, however, the intermittent nature of the process can result in inadequate fusion.

Gas shielding

The gas used for shielding in gas metal-arc welding must be of a suitable composition and purity, it must be delivered in the appropriate quantity and in an efficient manner.

Gas composition. For tungsten-arc welding the choice of gas is limited by the requirement that the tungsten electrode must not be affected. Only inert gases or nitrogen or their mixtures, sometimes with hydrogen, are used. Argon is most generally employed and can be used for any metal. For thick materials, particularly those of high thermal diffusivity, there is an advantage in using helium as this gas gives a higher arc voltage and deeper penetration. Mixtures of inert gases are also used.

An oxygen-free nitrogen shield can be used for copper with which it is inert. The use of diatomic gas increases the heat transferred to the plate, partly because of the concentration of the arc core and partly because of dissociation and recombination at the surface of the work of the diatomic gas. This extra heat with its associated deeper penetration is an advantage when welding high thermal conductivity metal such as copper. In welding nickel an addition of hydrogen to the inert gas has a similar effect and raises the arc voltage noticeably. Its reducing effect is also beneficial and it has been shown to be absorbed and to pass through weld metal to emerge on the underside where it protects from oxidation. For most other metals hydrogen must be excluded because of its deleterious metallurgical effects. The chief impurities in an inert gas are water vapour, oxygen and nitrogen, and these must be at a minimum when welding the reactive non-ferrous metals. A purity of at least 99·95 per cent with a dew point better than $-30\,°C$ is required.

With the gas metal-arc process there is more scope in the choice of the shielding gas because the electrode wire is transferred to the weld pool, and even where changes in composition occur due to gas–metal reactions these can often be rectified by the appropriate choice of electrode wire. Although used initially with argon or helium the process has also been used with nitrogen, carbon dioxide and mixtures of these with the inert gases. The shielding gas affects the burn-off rate, type of metal transfer and penetration. Gases are selected for use on the basis of these three parameters and economy.

Argon is generally used for welding non-ferrous metals, but mixtures of argon and helium are sometimes used where helium is obtainable. A compromise in arc properties is obtained with such mixtures. High current arcs in argon tend to show 'fingering' as shown in fig. 88 (p. 164), but this effect is less pronounced with helium which gives a broader penetration but a tendency to spatter. For ferrous metals an addition of oxygen is usually made to the inert gas. With additions of up to 5 per cent oxygen the transfer of metal is smoother and more regular with the end of the electrode pointed and enveloped by the arc giving axial transfer. It is the effect on the cathode or the work, however, which is of most importance. The fluidity of the weld pool and the ease with which it 'wets' the work is improved because of the presence of traces of the easily melted and fluid FeO. Bead shapes are therefore flatter and smoother. Even such small oxygen additions require the use of a deoxidizing electrode wire.

Where adjustments can be made to weld pool chemistry and the electrode or base metal does not form refractory oxides it is possible to use an active diatomic shielding gas. These gases are cheaper than the inert gases but they can also have technical advantages. Mixtures of argon and nitrogen have been used for copper to take advantage of the higher heat input rate with nitrogen and the favourable transfer characteristics with argon. The most used diatomic gas is carbon dioxide employed for welding mild and alloy steels. This may be employed alone or in mixtures with argon and oxygen. When used alone carbon dioxide gives a deep bowl-shaped penetration. This shape of penetration is preferred to the narrow, fingered penetration characteristic of argon-rich shielding gases as it is difficult to ensure that such narrow penetration is placed where required. The cross-sectional area melted is about 50 per cent greater than with the argon–oxygen mixture for the same welding current. Provided that the electrode wire contains sufficient deoxidizers such as silicon, manganese or aluminium, sound weld beads are formed in steel in a carbon dioxide atmosphere. About half the silicon and manganese present in the wire are used up in reacting with FeO to suppress the reaction $FeO + C \rightleftharpoons Fe + CO$ which causes porosity. The deoxidation products are left behind as fine inclusions in the weld metal and in the traces of slag left behind on the weld surface.

Even where only an inert gas is used it is frequently necessary to use electrode or filler wires which contain additions of deoxidizers, for example when welding copper and nickel base alloys. These additions are required because in practical welding there is a risk that shielding may not always be complete so that some oxidation can occur. There are also traces of oxides on the surfaces to be welded left from incomplete precleaning.

Gas flow. A gas shield is broken down by the entrainment of air from the surroundings. When a column of a gas is discharged into the air this entrainment is progressive, becoming deeper as the distance from the orifice increases. This is illustrated in fig. 41*b* for a free jet and in fig. 41 *a* for the situation appropriate to welding when an interface is present a short distance from the nozzle. The coverage afforded by a stream of a particular

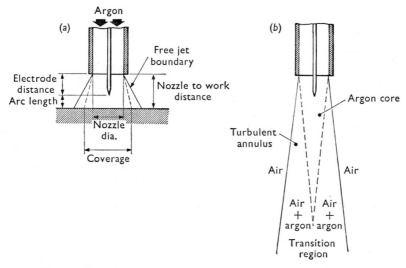

Fig. 41. Effect of entrainment on gas coverage for gas-shielded welding.

gas would appear, therefore, to be determined by the nozzle diameter and the nozzle/work distance. For maximum coverage, however, the gas must also issue from the nozzle in a non-turbulent manner.

Flow within a tube can be described by the relationship

$$R = Dv\rho/\mu,$$

where R is known as the Reynold's number, D is the diameter of the tube, v the average velocity, ρ the gas density and μ the viscosity of the gas. If the value of R exceeds 2300, which figure depends to some extent on the way the gas is introduced to the tube, the flow within the tube becomes turbulent. Under these conditions the coverage afforded by the gas stream from a nozzle would be greatly reduced. Coverage depends therefore on nozzle/work distance, nozzle diameter, gas velocity, the physical properties of the gas, and the design of nozzle and torch body.

The nozzle/work distance should clearly be held to a minimum consistent with the risk of overheating and collecting spatter and the need to allow visibility of the weld pool. For short nozzle/work distances the critical conditions leading to turbulence occur at higher Reynolds numbers than for the free jet. For a given nozzle diameter coverage is improved as the flow is increased up to that critical flow where turbulence begins to occur. As turbulence occurs at lower critical gas velocities when the nozzle diameter is increased there is a limit to the improvement in coverage which can be afforded by using larger nozzles. Taking R as 2300 the critical flow for argon through a 0·5 in. diameter nozzle would be approximately 40 ft^3/h and for a 0·75 in. diameter nozzle 60 ft^3/h. The critical argon gas velocities for these two nozzles are 475 and 317 ft/min respectively. With a dense gas such as carbon dioxide the conditions of turbulence are reached at lower gas velocities, the relative flow rates for CO_2 compared with argon being in the ratio of 0·6:1 approximately.

The efficiency of the shielding depends in large measure upon the design of the welding torch and the nozzle through which the gas issues to cover the work. A non-turbulent, laminar flow condition is achieved most easily if the torch has a bore which is parallel, smooth and long in proportion to its diameter. All torches have a diffuser of some type at the end of the bore where the gas enters so that gas does not blow directly down the barrel. The nozzle itself is best cut square and even a small bevel or radius on the inside edge reduces shielding efficiency. The effect of an arc in the gas stream is to heat the gas and increase turbulence at low flow rates, but with normal flow rates and welding currents the arc stabilizes the flow and improves coverage. With high current arcs the entrainment effect of plasma streaming must be considered and both nozzle diameter and flow rate should be enough to ensure that the entrained gas is argon and not air.

If gas-shielded welding is used in draughty conditions it is important that the gas velocity should be sufficient to give a stable flow. Similar observations apply when welding at high speed when the shielding gas is dragged behind the moving torch. The customary forward-pointing angle of the torch helps to overcome drag effects but where possible it is preferable to move the work rather than the torch.

Typical nozzles and flow rates for argon tungsten-arc welding of aluminium would be 10–20 ft^3/h for nozzles from $\frac{3}{8}$ to $\frac{5}{8}$ in. diameter bore. Inert gas metal-arc welding requires high flows of 40–60 ft^3/h for nozzles of $\frac{1}{2}$–$\frac{3}{4}$ in. diameter bore. For ferrous metals a lower flow rate may be used.

Protection from oxidation is not normally required for any part of the weld other than the weld pool and metal immediately adjacent to the weld pool. With highly reactive metals such as titanium and zirconium, however, protection must be provided for solid metal at elevated temperature. The ultimate in protection is provided by welding within a chamber or glovebox, but for titanium manual welding is possible in the open by using gas-backing techniques and a trailing supplementary argon shield on the torch.

Bibliography

Agnew, S. A. and Canulette, W. N. (1964). Improvement of gas metal-arc spot welds. *Weld. J.* **43**, no. 4, 184S–92S.

Amson, J. C. and Salter, G. R. (1963). Analysis of the gas shielded consumable metal-arc welding system. *Br. Weld. J.* **10**, no. 9, 372–83.

Borland, J. C. and Hull, W. G. (1958). Manual open-air welding of reactive metals. *Br. Weld. J.* **5**, no. 9, 427–34.

Burgess, N. T. (1961). Automatic tungsten-arc welding of heat exchangers. *Br. Weld. J.* **8**, no. 4, 141–50.

B.W.R.A., Cambridge, England (1956). *Argon-arc Welding Aluminium. Part 1. Principles and Bibliography. Part 2. Electrical Characteristics and Equipment for Argon-arc Welding Aluminium.*

Campbell, R. J. and Miller, D. R. (1961). *Weld. J.* **40**, no. 8, 828–38.

Cooper, G., Palermo, J. and Browning, J. A. (1965). Recent developments in plasma welding. *Weld. J.* **44**, no. 4, 268–76.

Copelston, F. W. and Gourd, L. M. (1958). Tungsten-arc spot welding. *Br. Weld. J.* **5**, no. 9, 394–9.

Gibson, G. J. (1953). Gas-flow requirements for inert gas-shielded arc welding. *Weld. J.* **32**, no. 4, 198S–208S.

Goldman, K. (1963). Electric arcs in argon. *Br. Weld. J.* **10**, no. 7, 343–7.

Griest, F. J. and Hawkins, R. L. (1964). Gas tungsten-arc welding technique with a new electronically controlled power supply. *Weld. J.* **43**, no. 7, 598–604.

Hackman, R. L. and Manz, A. F. (1964). D.c. welding power sources for gas-shielded metal-arc welding. *Weld. Res. Counc. Bull.* no. 97.

Kenyon, D. M. (1963/4). Arc behaviour and its effect on the tungsten-arc welding of magnesium. *J. Inst. Metals*, **92**, 9–13.

Method and apparatus for welding with gas shields having laminar flow (1951). Brit. Pat. 684,011.

Needham, J. C. and Carter, A. W. (1965). Material transfer characteristics with pulsed current. *Br. Weld. J.* **12**, no. 5, 229–41.

Needham, J. C. and Hull, W. G. (1954). Self-adjusting welding arcs. *Br. Weld. J.* **1**, no. 2, 71–7.

Needham, J. C. and Smith, A. A. (1958). Arc and bead characteristics of the aluminium arc in argon. *Br. Weld. J.* **5**, no. 2, 66–76.

Patriarca, P. and Slaughter, G. M. (1953). Cone-arc welding. *Weld. J.* **32**, no. 7, 579–602.

Rothschild, G. R. and Lesnewich, A. (1961). Inert-gas shielded arc welding of ferrous metals. *Welding Res. Counc. Bull.* no. 70.

Rowlands, E. W. (Jr.) and Cooksey, J. C. (1960). Internal welding of tubes to tube sheets. *Weld. J.* **39**, no. 7, 704–10.

Sekiguchi, H. (1962). Recent advancement of the CO_2–O_2 arc welding process in Japan. Doc. XII-127-62 Int. Inst. Welding.

Smith, A. A. (1965). *CO_2 Welding.* B.W.R.A.

Smith, A. A. and Dye, S. A. (1963). Significance of wire diameter in the CO_2 welding of pipe. *Br. Weld. J.* **10**, no. 5, 258–65.

Smith, A. A. and Poley, J. G. (1959). Effect of wire diameter on metal transfer (in the aluminium self-adjusting arc). *Br. Weld. J.* **6**, no. 12, 565–8.

Thielsch, H. and Pulliam, C. S. (1955). Engineering aspects of inert-gas tungsten-arc welding of piping. *Weld. J.* **34**, no. 7, 1185–95.

Tomlinson, J. E. and Slater, D. (1958). Automatic welding of aluminium plate. *Br. Weld. J.* **5**, no. 8, 361–8.

Winsor, L. P. and Turk, R. R. (1957). A comparative study of thoriated, zirconiated and pure tungsten electrodes. *Weld. J.* **36**, no. 3, 113S–19S.

5

UNSHIELDED AND SHORT-TIME ARC PROCESSES

In the early forms of bare-wire metal-arc welding and of carbon-arc welding no attempts were made to provide shielding from the atmosphere. The advantages of excluding air from contact with the arc and weld pool were quickly appreciated so that when arc welding is employed nowadays a flux or gas shield is invariably used. Attempts have been made recently, however, to render shielding unnecessary by correcting the consequences of atmospheric contamination. The method involves the use of gas metal-arc equipment, but without the gas. This development is entirely logical and should not be considered as a return to the old practice of bare-wire welding, as it makes use of much of the modern technology of gas metal-arc welding with continuously fed fine wires. Shielding is unnecessary because additions of deoxidizers—silicon, manganese, aluminium and cerium (see analysis in table 6)—are made to the wire. The cerium is reported to be beneficial in reducing the incidence of porosity due to nitrogen. See also flux-cored electrodes on p. 46.

Table 6. *Chemical composition of wire used for unshielded welding of mild steel (per cent)*

C	Mn	Si	Al	Ti	Ce	S	P
0·20	1·12	0·80	0·45	0·17	0·10	0·008	0·012

The remaining unshielded arc processes are short-time and discontinuous, and shielding is not normally required because either the whole process is carried out so rapidly that contamination is negligible or the molten metal is squeezed out of the joint. Processes in this group are percussion welding and arc stud welding. A common feature of the processes is that they permit the joining in a single event of parts with small cross-sectional area to other similar parts or more massive pieces, for example stud welding. A short-time electric arc is used to melt the end of the smaller part and form a molten pool in the surface to which it is to be joined. When the parts are subsequently brought together rapidly a weld is formed. Although for most applications no shielding is required it has been found that with aluminium alloys improved results are obtained with an argon shield.

The division between gas-shielded and unshielded arc processes tends to disappear, therefore, in the extreme case of welding metals forming refractory oxides.

PERCUSSION WELDING

The process appears in several forms but in every variation a short-time high-intensity arc is formed by the sudden release of energy stored generally, but not invariably, in capacitors. Subsequent very rapid or 'percussive' impacting of the workpieces to form the weld is required. The difference should be noted between this process and the capacitor discharge resistance welding method in which a capacitor is discharged into the primary of a transformer and heat is generated by resistance and not an arc (see p. 119).

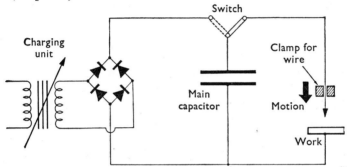

Fig. 42. Basic circuit for capacitor discharge or percussion welding.

Arc initiation

The main point of difference between the variations of percussion welding is in the method of initiating the arc. Three methods are used: (1) low voltage with drawn arc; (2) high-voltage breakdown; (3) ionization by a fusing tip. With each method the energy source can be a bank of capacitors, possibly as part of delay line, which is charged by a variable voltage transformer/rectifier unit. Both charging voltage and capacity are variable so that the energy available and the shape of the current–time curve can be adjusted. The simplest circuit arrangement is illustrated in fig. 42.

Low-voltage, drawn arc. The earliest percussion welding process was devised by L. W. Chubb in the U.S.A. in 1912. A low-voltage capacitor was used, the workpieces were initially in contact but were separated by an electro-magnetic device as the discharge commenced. When the arc had melted the surfaces to be joined they were brought together under impact.

High-voltage breakdown of gap. It was not until 1941 that percussion welding was seriously employed, this time by F. C. Vang, who eliminated the mechanism for drawing the arc by using energy stored at high voltage. In this method the parts to be joined are connected to the capacitor and then forced together by a spring or by compressed air. As the gap between the parts closes to a distance which the voltage can break down, an arc of high intensity is initiated and superficial melting takes place before the parts impact together. Breakdown in this manner will only occur with voltages of the order 10^3 V and since these are lethal the equipment using the process is usually automatic. An advantage of the high voltage is that considerable power may be conducted down a fine wire to make a weld at its end, without the risk of fusing it. Welds can be made rapidly with a corresponding benefit from minimal heat spread.

This type of process has been widely used for connecting wires to electrical equipment, the wire frequently being in coil form fed automatically and cropped to length after the weld has been made. The weld time is short and the equipment operates at a high repetition rate so that its dynamic design is important, particularly with regard to bounce in the working parts. Conditions for welding wire of say 0·040 in. diameter might be a charge voltage of 1 000+ V with an approach velocity of 40 in./s and an arc duration of less than 100 μs.

Low-voltage, ionizing tip. The third method of using energy stored in a capacitor employs low voltage and ignites the arc by the fusing of the initial point of contact between the workpieces. Metal is vaporized which provides an ionized path for the power arc. Two well-developed commercial applications have become generally known as spark discharge welding and capacitor discharge or percussion stud welding. The former for joining fine wires such as thermocouples or lead wires to surfaces, and the latter for welding studs and fasteners to sheet material.

Spark discharge welding

For joining wires to more massive parts a simple capacitor bank charged with voltages of up to about 100 V is used, the voltage being applied to the wire before contact is made with the workpiece. A dabbing action, either manual or mechanical, is used, and on first contact a short length at the tip of the wire is vaporized or splashed out by the initial surge of current. The arc formed in the gap which is left melts a globule on the end of the wire and a pool on the workpiece. Follow through of the motion results in the gap closing with decrease in voltage until the wire is plunged into the pool. The decrease in circuit resistance when this occurs causes a second surge in current as the remaining charge on the capacitors is dissipated. The duration of the arcing period is usually only several milliseconds. All percussive welding processes show the same pattern of growth and decay of current, however, when $R^2 < 4L/C$, the discharge is oscillatory. The discharge time is proportional to capacitance and the energy stored to the square of the voltage.

Although metallographic sections of the joints may not give an impression of high quality, joint strengths are good with failure during testing frequently taking place in the wire. As indicated in fig. 43 the wire is embedded in a crater of solidified weld metal, the edge of the crater usually being surrounded by spatter produced during the arcing process. Dissimilar metals are often joined by spark discharge welding and the resultant weld metal if tested in bulk would often be brittle. The thickness of weld metal between the wire end and the base is thin, however, so that even brittle weld metal compositions can behave satisfactorily in mechanical testing. To obtain consistent results attention should be given to the control of the percussive motion and to the condition of the end of the wire. Clean wire cut so as to give a slightly pointed end behaves well, although even wires with ball ends can be welded satisfactorily if the end is oxide free and a higher capacity setting is used.

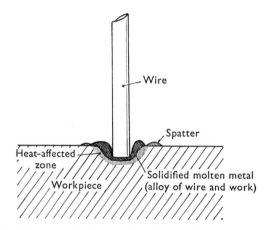

Fig. 43. Diagrammatic cross-section of spark discharge wire weld.

For certain applications a spring-loaded wire initially in contact with the workpieces is preferred. When this is done a mercury switch is used in the lead from the capacitor; however, the arc initiation is performed in the same manner as described above. To weld two wires together or to fuse small assemblies the capacitor may be used to provide an arc from the wires to a refractory electrode such as carbon or tungsten, if necessary with a gas shield.

Capacitor discharge or percussive stud welding

An ionizing tip is also used in the stud welding version of percussion welding. The process generally involves welding a stud or fastener of up to 0·25 in. diameter, to relatively thin sheet metal. Arc initiation is by the vaporizing of a small cylindrical tip which projects from the centre of the stud (fig. 44*a*). Reliable starting requires consistency in the shape and condition of this tip. This part of the process takes place in less than 1 ms and peak power is reached within this time with currents of the order 10^3 A. Subsequently, the anode and cathode roots of the arc move rapidly over the surface of the stud and workpiece to cause superficial fusion of both. A period of less than 10 ms is allowed for this part of the process during which time the current is falling as the capacitors are dis-

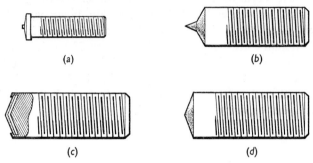

(a)   (b)

(c)   (d)

Fig. 44. Types of stud for capacitor discharge and arc stud welding: (a) small cylindrical pip for capacitor discharge; (b) flux sprayed double cone on aluminium alloy stud; (c) steel stud with end cap filled with granular deoxidants; (d) single cone steel stud sprayed with aluminium.

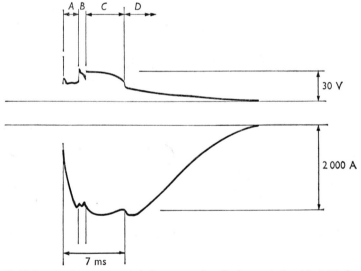

Fig. 45. Voltage and current records for a capacitor discharge stud weld. A, Pip heating; B, pip fusing; C, arcing period; D, weld complete, charge decaying. *Note:* voltage measured across stud and work. Current oscillogram inverted during recording.

charged and the gap is closing as the stud is moved forward under the percussive force. When contact is made the melted regions combine and molten metal is spread between the joint faces. Figure 45 shows the basic wave-forms of voltage and current for the sequence described and plate 6 the form of the joint.

The layer of weld metal between stud and sheet often contains small voids possibly because of incomplete excursion of the arc over the joint faces. Because the stud invariably has a large diameter compared with the thickness of the sheet, however, the joint strength is adequate and failures in testing invariably occur in the sheet metal. As with spark discharge wire welding the use of dissimilar stud and sheet material is common and, although some of the combinations would be quite unacceptable in massive form, the design of the joint and limited heating result in satisfactory joints. The depth of penetration into the workpiece is 0·005–0·010 in. and the thickness of weld metal between stud and sheet is approximately the same.

Techniques and equipment. Two methods of starting the process are in use known as 'initial contact' or 'initial gap' according to whether or not the tip of the stud is in contact with or away from the sheet. In the former method the stud which is held in a spring-loaded plunger has its tip pressed against the workpiece. On closing a control switch the capacitors are connected to the plunger, the stud tip vaporizes and the process begins. With the 'initial gap' method the stud is held about $\frac{1}{16}$ in. away from the work against the spring by a solenoid surrounding the plunger. To make a weld the capacitors are switched in, the solenoid is de-energized and the stud moves forward, the discharge being initiated when the tip makes contact with the work. This method is used when fastenings must be made to light gauge aluminium or where minimum spread of heat must take place to the opposite side of the sheet, for example when welding sheet painted or anodized on the reverse side. Because the stud is already moving when the discharge begins the process is effectively more rapid. Care is taken in the design and maintenance of the welding tool to ensure smooth, rapid and consistent motion of the plunger. The absence of sliding surfaces which is achieved by mounting the plunger on diaphragms is an advantage.

A circuit for use with the tool described is illustrated in fig. 46. An important feature of this circuit is the use of an ignitron rather than an electro-mechanical switch. Ignitrons are rapid in action and do not have the disadvantage of contactor bounce associated with the electro-mechanical switch. The operation of the switch on the hand tool opens the multiple contacts S_3, disconnecting the capacitors from the charging circuit and discharging capacitor C_6 which fires the ignitron to complete the power circuit. The time constant of the circuit, controlling the rate at which the capacitors are discharged, can be influenced by variations in the inductance of the welding leads. To minimize this effect the inductance of the leads is made small in comparison with the total by inserting an inductance L in the equipment.

In the diagram unit *VCS* is a voltage control and sensing device. With the simple circuit shown in fig. 42 (p. 95) the voltage across the capacitor rises exponentially to the transformer voltage. The last few volts across the capacitor are, therefore, achieved relatively slowly and the final voltage can be influenced by mains fluctuations. The voltage control and sensing unit in fig. 46 is always set to maintain a voltage less than that available from the rectifiers and disconnects the charging unit when the pre-set voltage has been reached across the capacitors. Because of the exponential voltage build-up this voltage is achieved more rapidly than with the simple circuit. If the capacitor voltage drops through leakage as a result of delay in making a weld the unit will re-connect the charging supply.

Although only portable manually operated equipment has been described machine-type welding heads are also used with appropriate tooling. These units employ the initial gap principle and use compressed air for the application of the percussive force. Although the arcing period is short arc blow can still occur and twin earth return leads are frequently used to minimize this trouble.

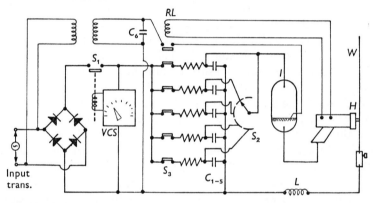

Fig. 46. Circuit for manual capacitor discharge stud welder. C_{1-5}, Main capacitors; C_6, triggering capacitor; *VCS*, voltage control and sensing unit; S_1, voltage control switch; S_2, capacitor selector switch (current control); S_3, multi-relay; *RL*, firing relay; *I*, ignitron; *L*, inductance; *H*, handtool; *W*, work.

Applications. Percussive stud welding is used for welding studs, fastenings and other small fittings to domestic, electrical and architectural products. The workpiece may be sheet metal, often coated, painted or polished on the reverse side or a die-cast or sintered component. Many dissimilar metal combinations can be used including mild steel studs to zinc-base die castings. Generally shielding is unnecessary, but for aluminium, which forms a refractory oxide, improved results are obtained with argon shielding.

Percussive welding with magnetic force

A percussive process widely used for welding electrical contacts to backing strips has features in common with resistance welding. The ionizing tip which is formed in the backing is a projection of similar design to that used in projection welding. Current is supplied not by capacitors but by a transformer with an open circuit voltage of 10–20 V which is higher than is normal for resistance welding. The current is sufficiently high to vaporize the projection so that the normal process of collapse does not occur and a short time arc is formed, as in flash welding, with violent expulsion of metal. A percussive force is applied by the motion of an armature within the secondary circuit of the transformer which is responsive to the welding current. (See also magnetic force spot welding.) This force closes the gap between the workpiece with explosive violence, trapping much of the splash from the projection. Applications for this process are similar to those for percussive stud welding, particularly the welding of dissimilar metals. As the operation must be done in a press-type welding machine, however, the element of portability is lacking.

ARC STUD WELDING

Joints are frequently required between the ends of rods or bars and a metal surface such as plate or section. One method of making this type of joint is to draw an arc between the rod and the surface and then bring the molten surfaces in contact. The simplest joint of this type replaces the stud which is drilled and tapped into a surface and the process is therefore known as stud welding. Arc stud welding, invented by H. Martin and used from 1918 by the Royal Navy was not widely known until it was redis-covered by E. Nelson in the U.S.A. twenty years later. Almost twenty years later still W. P. van den Blink and others devised the stud-welding process in which the arc is ignited by a fusible collar and not by drawing the stud away from the workpiece as in the previous methods. All three methods are now in use side by side. Apart from differences in the way in which the arc is initiated the three variations of stud welding differ in the mechanism by which the stud is returned to the workpiece. In compari-son with the percussion stud-welding process the arc method employs arc times of between 150 and 500 ms, twenty-five times as long, while the motion of the stud is greater. The longer arc times and less critical opera-tion allow arc stud welding to be used with studs up to $\frac{7}{8}$ in. diameter.

Drawn arc method

Hot-plunge process. Basically the original method invented by Martin. The stud is held initially in contact with the workpiece by a spring. To weld the stud a solenoid is energized which lifts the stud away from the work a pre-set distance against the action of the spring. With a pilot current of 15 A an arc is drawn and when the stud has lifted the pre-set amount of about $\frac{1}{16}$ in. the welding current is switched in to form the power arc. The welding current is supplied from a drooping characteristic d.c. power source, usually a transformer/rectifier. When the current has passed for a pre-set time (long enough to melt the end of the stud and a crater in the workpiece) the solenoid is de-energized and the spring returns the stud to the work. This is done while the current is still flowing, hence the name hot-plunge method. This sequence is illustrated in fig. 47. The decrease in circuit resistance when the stud meets the work causes an increase in the current flowing of up to 100 per cent and this current flows until the main power contactor cuts out. A current surge of this magnitude can cause overheating and limit the life of the contactor so it can be arranged for an auxiliary contactor to switch in a resistance at the end of the cycle to limit the current

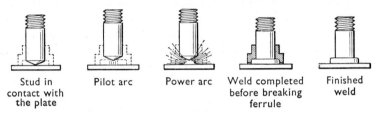

| Stud in contact with the plate | Pilot arc | Power arc | Weld completed before breaking ferrule | Finished weld |

Fig. 47. Stages in the arc stud weld process.

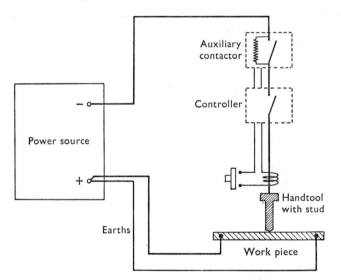

Fig. 48. Simplified circuit for arc stud welding.

rise (see fig. 48). The fact that the current flows until after the weld has been made is of significance when welding high thermal conductivity metals such as aluminium.

Cold-plunge process. Devised by Nelson. Once more the stud is held in contact with the work by a spring and retracted by a solenoid. This solenoid, however, is in the main power circuit so that immediately the welding current is switched on the stud lifts drawing an arc on full power. When the welding contactor is opened after the pre-set time the arc is extinguished, the solenoid is de-energized and the stud plunges into the crater. Although this simple system is satisfactory for steel the time while the stud is being returned to the work in which no heat is generated is sufficient to make the process unsuitable for aluminium alloys. The deep initial crater formed by the full power arc also makes the method difficult to use on very thin metal. For a wide range of work on mild steel, however, the simplicity of the method and the rugged equipment is considered an advantage.

Controlling fillet shape. An improvement in the reliability of both the methods described so far resulted from the use of a ceramic ferrule placed over the end of the stud. This ferrule is shaped inside so that when the stud is plunged into the weld pool the weld metal which is expelled is formed into a smooth fillet round the base of the stud. The ferrule has a serrated or slotted lip so that the arc region is vented. Apart from shaping the fillet, the ferrule shields the operator from the glare of the arc and when pre-placed in jigs helps the positioning of the studs. Ferrules are expendable and are broken off the stud after the weld has been made. The form of the weld is indicated in the photomacrograph cross-section (pl. 7).

Metallurgical controls. Much of the molten metal which is necessarily exposed to the atmosphere in making the weld is eventually expelled into the fillet round the stud. Welds with steel studs, however, may have inferior mechanical properties or exhibit porosity if attempts are not made to correct the effects of contamination by oxygen and nitrogen. With the 'hot plunge' process this deoxidation is provided by aluminium sprayed on the conical end face of the stud (fig. 44 d). In the 'cold plunge' process a cap is fitted to the end of the stud which encloses a small quantity of powdered metal with high silicon, manganese and aluminium contents (fig. 44 c).

Aluminium studs have their ends sprayed with metallic zinc which it has been found reduces the incidence of porosity. The mechanism of this improvement is not understood but it may be due to the combined effects of preventing the end-face from acquiring a hydrated oxide film during storage and the flushing of the arc region with zinc vapour.

Pre-heating, provided by a low current 70 A arc before the full welding current is switched in, is used on both ferrous and aluminium alloy studs. With the former it may be used to delay the cooling rate of stud welds in hardenable materials while with the latter it provides a more gradual temperature gradient in the stud, ensuring good fusion between the stud and the fillet of diplaced molten metal. Good fusion between fillet and stud is also dependent on the surface of the stud being oxide free. Aluminium studs are therefore machined to remove the mill surface and may be welded in an argon stream to improve strength and consistency.

Fusible collar method

A collar as shown in fig. 49 having a composition similar to the contact metal-arc electrode covering and having a cardboard cap is placed on the end of the stud. Stud and collar are then held against the workpiece by a spring-loaded hand tool. On completing the circuit through a contactor the collar conducts sufficient current to heat up and provide initially an ionized path for the arc from stud to work. Part of the collar fuses, forming a protective slag over the pool, and finally, after the arc has been running for a time depending on the size of the collar, the seating gives way under the spring pressure, allowing the stud to be driven into the molten pool. The expelled metal forms a smooth fillet round the stud under the envelope

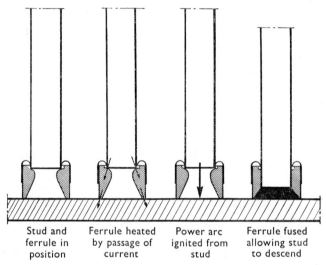

| Stud and ferrule in position | Ferrule heated by passage of current | Power arc ignited from stud | Ferrule fused allowing stud to descend |

Fig. 49. Stages in the fusible collar method of arc stud welding. Left: ready to weld; right: shoulder in collar collapses allowing stud to plunge into weld pool. The collar is enclosed in a cardboard cap.

of slag. Because the collar is responsible for both the initiation of the arc and the timing of the operation the equipment for this process is extremely simple. By virtue of the arc stabilization by the collar the fusible collar process unlike the drawn arc methods can be used with an a.c. supply. Normal drooping characteristic power sources with open circuit voltages of 70–100 V are used, and although a.c. is preferred for cheapness the process works satisfactorily on d.c. with the stud positive.

Square-ended untreated studs are used because the slag formed by the fusing of the collar provides both atmosphere control and deoxidation of the molten metal. The size of the collar must be matched to the diameter of the stud being used otherwise faulty timing of the operation results. Being dependent on a metal-arc-type electrode covering composition for the collar restricts the process to mild and alloy steels and some stainless steels. There is also the geometric limitation which prevents the use of studs with non-circular cross-sections.

APPLICATIONS FOR ARC STUD WELDING

Important uses of arc stud welding are in civil engineering for shear connectors in ferro-concrete constructions, in shipbuilding and general engineering for insulation attachment, deck laying, cable cleats, cover and panel fixing, extended surfaces on heat exchangers and for fixings of all types. With suitable selection of process studs may be welded from $\frac{1}{8}$ to 1 in. diameter, or equivalent area, in steel, stainless steel, brass, or aluminium alloy. No elaborate surface preparation of the work is necessary, although more consistent results are obtained with shot-blasted, lightly ground or wire-brushed surfaces.

Bibliography

Blink, W. P. van den, Ettema, E. H. and Willigen, P. C. van der (1954). A new process of stud welding. *Br. Weld. J.* **1**, no. 10, 447–54; discussion (1955), **2**, no. 3, 113–15.

Laurie, D. J. N. (1961). Developments in stud welding. *Br. Weld. J.* **8**, no. 4, 113–21.

Manning, R. F. and Welch, J. B. (1960). Percussion welding using magnetic force. *Weld. J.* **39**, no. 9, 903–7.

Moravskii, V. E. and Kaleko, D. M. (1964). Condenser-discharge percussion welding of components made of substances of high electrical conductivity. *Automatic Welding*, no. 3, 13–18. (B.W.R.A. translation.)

Product Profile (1961). *Engineering*, pp. 786–7.

Quinlan, A. L. (1955). Automatic percussion welding of telephone relay contacts. *Weld. J.* **34**, no. 3, 237–40.

Slutskaya, T. M., Krivenko, L. F., Avramenko, V. A. and Kovalev, Yu. Ya. (1963). An electrode wire for mechanized welding of steel with no shielding medium. *Automatic Welding*, no. 8, 19–25. (B.W.R.A. translation Avt. Svarka.)

Sumner, E. E. (1955). Some fundamental problems in percussive welding. *Bell System Tech. J.* **33**, no. 4, 885–95.

6

RESISTANCE WELDING

Two ways exist of utilizing an electric current to produce heat directly in a metal. The current may be used to maintain an arc to the surface of the workpiece, as in arc welding, or heat may be liberated by the passage of the current through the work. In this latter method heat is generated by the resistance to the passage of the current according to Joule's law

$$H \text{ (cal)} = I^2 RT/J,$$

where J is the electrical equivalent of heat.

Heat in arc welding is generated at the surface and is distributed through the workpiece by conduction. In the resistance method heat can be liberated throughout the entire cross-section of the joint. The electric current which generates the heat may be introduced to the work through electrodes with which the work makes contact, or it may be induced within the metal by a fluctuating magnetic field which surrounds the work. Although both methods depend on resistance heating the term resistance welding is often used only for the former. The latter process is known as induction welding.

A variety of resistance-welding methods exist depending on the different ways of creating a locally high resistance so that heating may be concentrated at this point. Actual resistance depends on both the resistivity and geometry of the conductor. Since the resistivity is fixed by the workpiece materials it is usual to create the local high resistance by providing a restricted current path between the parts to be joined, a procedure known as current concentration. All resistance-welding methods require physical contact between the current-carrying electrodes and the parts to be joined. Pressure is also required to place the parts in contact and consolidate the joint and these are features which distinguish the processes from most arc welding methods.

RESISTANCE SPOT WELDING

This process, in which overlapping sheets are joined by local fusion caused by the concentration of current between cylindrical electrodes, came into use in the period 1900 to 1905. It is now the most widely used of resistance-welding processes. Figure 50 is a diagrammatic arrangement of the process. The work is clamped between the electrodes by pressure applied through levers, or by pneumatically operated pistons. On small welding machines springs may be used. Current is generally supplied by a step-down transformer, the work, electrodes and arms of the machine being part of a secondary circuit consisting of only one or two turns. A spot-welded joint comprises an array of one or more discrete fused areas or spots between the workpieces.

Electrode and nugget size

Current concentration is determined by the area of contact between the electrodes and the work, and clearly the size of the weld or nugget of fused metal is closely related to this area. The shear strength of the nugget must

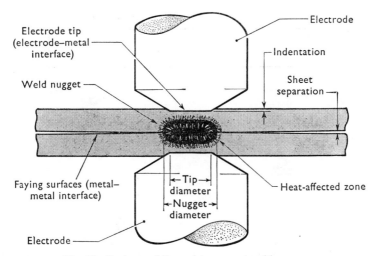

Fig. 50. Features of the resistance spot-weld process.

usually be sufficient to ensure that when the joint is stressed to failure this occurs in the sheet around the nugget. A desirable form of failure is the pulled slug type where fracture occurs in one sheet round the nugget which is left attached to the other sheet.

A method for standardizing the size of the electrode according to the sheet thickness has become accepted which has its origin in the procedure for riveting. Riveted joints are often designed on the basis $d = 1 \cdot 2 \sqrt{t}$, where d is the rivet diameter and t the sheet thickness. The relationship is probably of this form because the efficiency of a single-row riveted joint, rivet strength/sheet strength, is $(\pi d^2 f_s)/(4 p t f_t)$, where p is the pitch of the rivets and f_s and f_t are respectively the shear strength and tensile strength of the rivet and plate. For any given pitch d^2/t is constant. In a riveted joint the hole for the rivet weakens the plate. With a spot-welded joint the weld is integral with the plate so that the only weakening occurring is that due to the softening effect of the welding heat. Higher efficiencies can therefore be achieved in welding. Because weld diameter is so closely related to electrode diameter the latter may be quoted for spot welding and the electrode diameter d_e is made equal to $\sqrt{t}$. The above discussion should not give the impression that the accepted relationship, although useful, is in essence anything but empirical. Other formulae for relating electrode diameter to sheet thickness such as $d = (0.1 + 2t)$ are used which give substantially the same result except for very thin or thick metal where the $\sqrt{t}$ formula is more reasonable. For two sheets of 0.064 ($\frac{1}{16}$) in. thickness, for example,

$$d = \sqrt{t} = \sqrt{\tfrac{1}{16}} = \tfrac{1}{4}, \text{ giving an electrode of 0.25 in. diameter,}$$

or $d = (0.1 + 2t) = (0.1 + 0.128)$, an electrode of 0.228 in. diameter.

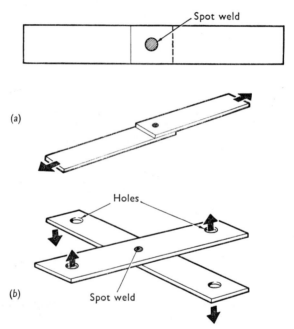

Fig. 51. Two of a number of mechanical tests for resistance spot welds:
(*a*) the shear strength test; (*b*) the cross-tension test.

It is important to recognize when considering the strength of a spot weld that in sheet metal the weld is rarely if ever stressed solely in shear because of distortion which takes place round the weld under load. Under these conditions the ductility of the parent metal at the periphery of the weld can have a dominating influence. A useful indication of weld ductility is obtained by taking the ratio of cross-tension strength (f_t) and shear strength (f_s) (fig. 51). This ductility ratio $f_t/f_s \rightarrow 1$ for maximum ductility and approaches 0 when extreme brittleness is present. These considerations are particularly important when welding material susceptible to quench hardening because of the high cooling rates in resistance welding.

So far only the area of contact between electrode and work has been considered. In practice this cannot be controlled by using electrodes in the form of rod of the required diameter—such electrodes would be mechanically weak and have too high a resistance. Practical electrodes are made from copper or copper alloy bar of substantial diameter machined to a truncated cone with an angle of 30°. Alternatively, electrodes may be machined with a domed end, the radius of the dome being used to control the area of contact. Clearly, electrode load and sheet hardness are also

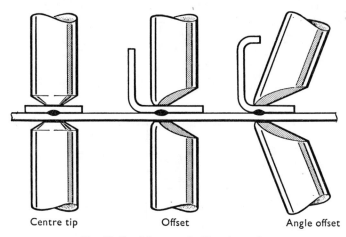

Centre tip   Offset   Angle offset

Fig. 52. Special types of offset electrodes.

significant factors in determining the area of contact with domed electrodes. Contact area is controlled more accurately with truncated cone electrodes and any wear in service can be readily seen. Compared with the domed electrode, however, they result in more obvious surface marking of the workpiece and require more accurate alignment. Although simple symmetrical electrodes are preferred a variety of special shapes is used to obtain access with complicated joints (fig. 52).

Resistance and force

Having established the relationship between electrode shape and workpiece thickness it is possible to examine the part played by the three important process variables, electrode clamping force, current and time of current flow. When the parts to be welded are clamped between the electrodes the inter-electrode resistance comprises five separate resistances, as shown in fig. 53. Of these five resistances where heat can be developed number 3, the interfacial or sheet/sheet contact resistance, is the most important as it is at this position that the nugget and, therefore, the heat is required. Resistance here is important early in the weld period. Consistent weld size depends, among other things, on consistency of surface condition at this interface. With low resistivity metals resistances 1, 3 and 5 assume greater importance and the need for their control is increased. Aluminium alloys, for example, are subjected to a rigorous pre-weld treatment of degreasing and pickling a limited time before welding to ensure consistent

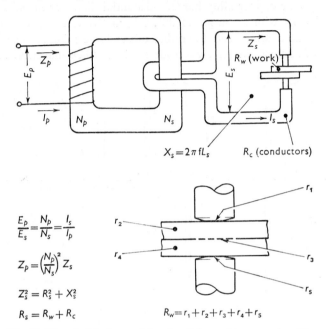

$$\frac{E_p}{E_s} = \frac{N_p}{N_s} = \frac{I_s}{I_p}$$

$$Z_p = \left(\frac{N_p}{N_s}\right)^2 Z_s$$

$$Z_s^2 = R_s^2 + X_s^2$$

$$R_s = R_w + R_c$$

$$R_w = r_1 + r_2 + r_3 + r_4 + r_5$$

Fig. 53. Electrical features of the circuit for resistance spot welding.

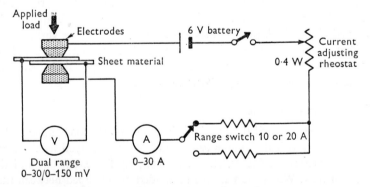

Fig. 54. Apparatus for measuring surface resistance.

contact resistance. Where such controls are required it is usual to check the contact resistance periodically by measuring the voltage drop when a current is passed across a pair of samples clamped together. An apparatus of the type shown in fig. 54 is used and resistance is calculated using

Ohm's law. Contact resistances are usually in the range 50–100 $\mu\Omega$ but may be 20 $\mu\Omega$ for aluminium.

The body resistances 2 and 4 depend on the resistivity and temperature of the work and cannot be altered. Body resistance has a major effect later in the weld period. Contact resistances 1 and 5, electrode to sheet, are wholly undesirable and kept to a minimum by using high conductivity electrodes and ensuring that there is adequate cleanliness and clamping force. Unfortunately, the requirements of high electrical and thermal conductivity are not compatible with good mechanical strength and wear

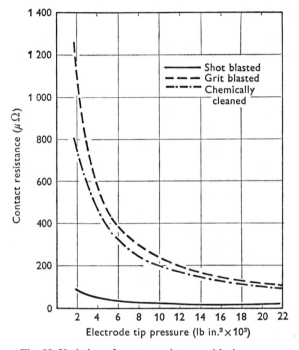

Fig. 55. Variation of contact resistance with tip pressure.

resistance at elevated temperatures. A variety of copper alloys, for example chromium–copper, cadmium–copper or beryllium–cobalt–copper, are employed which provide a range of properties suitable for different applications. The ill effects of resistance at the work surface, surface pick-up, splashing and electrode wear can be mitigated by water cooling the electrodes internally so that heat is conducted away rapidly. Efficient electrode cooling is essential with high production rates.

The effect of increasing electrode force is to reduce the contact resistance as shown for a heat-resisting steel in fig. 55, but the anomaly is noted that contact resistance is independent of area for any one electrode force. Contact resistance has been defined as comprising two parts; true contact resistance which is influenced by load, surface finish and condition and spreading resistance caused by the restriction of current through local points of contact. Spreading resistance is proportional to the resistivity of the material and the number of points of contact in parallel. These points of contact are possibly associated with the rupturing of surface oxide films where asperities are deformed. Contact resistance, depending on the resistances at the multiple points of contact which are deformed by pressure, does not return to its original value once the load is released.

For consistent operation high electrode forces would be desirable, but particularly when welding low resistivity metals the contact resistance cannot be reduced too far by raising force as it must be present to develop the welding heat. Excessively high forces are also undesirable because of the increase in surface indentation of the work and wear of the electrodes.

Force is maintained for several cycles after the current is cut off to consolidate the nugget.

Current and time

The effects of current and time can be considered together but, while they both affect the quantity of heat developed, it is the current alone which determines the rate of heat development. While the current is passing some of the heat generated is lost, mainly to the water-cooled electrodes. The size to which a nugget will grow, and indeed whether a nugget will form at all, depends on the heat being generated faster than it is removed by conduction. Current, therefore, is a most critical variable.

The relationship of current, time and thickness for welding two equal thicknesses of mild steel can be summarized in a manner similar to fig. 74 (see p. 133). When establishing procedures for welding a particular material and thickness, however, the strength/current curve at fixed times is most useful (fig. 56). Strength/time curves for fixed currents are similar. Each material has its characteristic curve—steeply rising curves with a sharp cut-off indicating that the setting of current is critical and, therefore, that the metal is difficult to weld.

Because of the selection of electrode sizes in proportion to sheet thickness a certain rationalization in choice of current and time is feasible. From an inspection of experimental and published data Humpage and Burford proposed simple formulae for determining current and time for mild steel. These were of the form:

(1) Current density $= 120000 + k e^{-25t}$ amp/in.^{2}, where t is the sheet thickness (in.) and k a constant which for present-day techniques approximates to 300000.

(2) Time $= 250t$ cycles (of 50 c/s supply).

The actual level of current required for any metal tends to be inversely proportional to its electrical and thermal resistivities. Copper is impossible to weld because the total resistance of the joint cannot be raised sufficiently above that of the secondary circuit of which it is a part. The insertion of a shim of high resistance low melting point alloy allows heat to be generated between the workpieces, but the process is then called resistance brazing. Electrodes having high electrical and thermal resistivities also assist by generating heat and restricting the conduction of heat away from the joint.

The above discussion assumes a constant size of electrode tip because current concentration is of equal importance with current. In use electrode tips wear and spread thereby reducing current density and weld size.

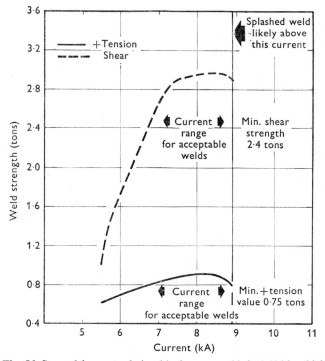

Fig. 56. Strength/current relationship for spot welds in 0·104 in. thick martensitic stainless steel; weld time 100 cycles.

Nugget formation

Further important aspects of the force, current and time relationship become apparent in tracing the growth of the weld. Resistance welds are characterized by their rapid formation and the steep heating and cooling curves. This is because of their local nature and the proximity of the electrodes.

Having applied the electrode force the passage of current is initiated and almost immediately, frequently within one cycle, there is a drastic drop in contact resistance. Because of the current concentration and resistances 1, 3 and 5 the temperature at the sheet/sheet interface and in two annular

8   HWP

regions under the electrodes rises rapidly. Although the contact resistance is lowered quickly, the heated metal in this region provides a locally higher resistance and the temperature at the interface continues to rise. Resistance welding could not be carried out if metals did not have positive temperature coefficients of resistance. Figure 57 shows the temperature distribution in a partly finished $2 \times \frac{1}{4}$ in. mild steel spot weld as determined by

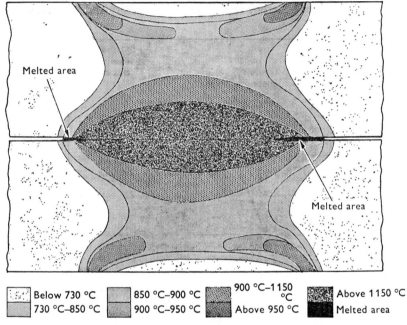

Below 730 °C	850 °C–900 °C	900 °C–1 150 °C
730 °C–850 °C	900 °C–950 °C	Above 950 °C
		Above 1 150 °C
		Melted area

Fig. 57. Temperature distribution in a mild steel spot weld during the formation of the nugget, as determined by metallographic examination.

metallographic means. As the process continues a molten nugget develops, the diameter of which increases rapidly at first and then more slowly as the maximum size is approached which may be up to 10 per cent greater than the electrode diameter. In a spot weld where the parts fit well and there is adequate electrode force, the molten nugget is safely imprisoned between the sheets, although it is subject to hydrostatic pressure from the electrodes. As the heat spreads in the later stages of the weld cycle the electrodes begin to sink into the work surface and as a result of plastic deformation the sheets begin to separate at the weld edge. These effects, indentation and sheet separation, set a top limit on current and time. The flow of current through the nugget causes turbulence in the liquid metal as can be seen from pl. 8.

The pressure and plastic deformation to which the ring of heated metal sealing the periphery of the nugget is subjected can result in corona bonding, when the workpieces are susceptible to pressure welding as with aluminium. Accidental rupture of this seal results in some of the molten nugget metal being spewed out between the sheets. This is called expulsion, and the weld is said to be 'splashed'. Expulsion puts a top limit on current and is promoted by low electrode force, bad fit or lack of mechanical support as a result of welding close to an edge. An edge distance of not less than $1\frac{1}{4}/1\frac{1}{2}d$ is desirable although narrower distances are used sometimes with reduced welding currents. It is also possible for expulsion to occur at the electrode/work interface, because of the too rapid or excessive generation of heat at the interface when scale, for example, has created a locally high resistance. Low resistivity metals are prone to surface splashing because electrode/sheet resistance is a higher proportion of the total resistance.

The process of heating the work and melting a nugget results in thermal expansion which, in spite of the electrode force, tends to separate the electrodes. Since this electrode separation is directly proportional to the heat liberated between the electrodes its measurement may be used as a means of quality control. Records of the more important welding parameters are shown in pl. 9.

Process and quality control

Resistance spot welding is an automatic process in which all process variables must be pre-set and maintained constant. This is necessary because at present once a weld has been initiated there is no way in which its progress can be controlled. Further, the non-destructive testing of the welds once they are made is difficult and not completely satisfactory. It is customary, therefore, to establish welding schedules by experiment and to maintain the best possible control of process variables in practice by making periodic destructive tests on either the product or test samples.

Regulating the welding current. Welding current or 'heat' as it is often loosely called is set on the machine either by changing the turns ratio of the transformer or by phase shift control. Assuming that the resistance and inductance of the secondary circuit do not change, the current flowing with a single-phase machine is proportional to the applied voltage. Current can be varied by adjusting the turns ratio through altering the tapping on the primary side of the transformer.

In phase shift control the transformer primary is supplied through ignitrons, a form of gaseous discharge valve, which act simply as high-speed switches. At the beginning of each half cycle of primary current the ignitron is made non-conducting, but later in the half cycle, after a delay which is adjustable, it is allowed to pass current. Current continues to pass until the current zero point is reached when the discharge stops. Only a portion of the available power therefore has been passed through to the secondary. On the following half cycle the procedure is repeated, but because ignitrons are also rectifiers and current is now passing in the opposite direction a second ignitron must be employed. The two ignitrons are connected in what is known as inverse-parallel or back-to-back as shown in the circuit (fig. 59).

The ignitrons themselves are each fired by a signal from a thyratron valve. Control of the voltage on the grid of this valve determines the point during each half cycle when it, and the associated ignitron, will pass current. This provides a convenient means of changing the power delivered to the secondary circuit. The resultant current wave-form has the appearance shown in fig. 58 and is called a chopped sine wave. Power is proportional to the sum of the areas under the remainder of the half-cycle current loops.

Fig. 58. Chopped sine wave current wave-form for a single-phase a.c. welder with heat control.

Power-source types. The electronic method of heat control simplifies switching and timing and also allows such operations as slope control. Slope control is a method of raising the current gradually to the welding value or of reducing it gradually instead of switching on and cutting off sharply at the beginning and end of the weld. In some areas where the supply voltage may fluctuate, voltage compensation devices can be fitted which hold the current constant in spite of line variations.

So far only single-phase a.c. power sources have been considered; however, this power source although the most important is only one of three possible systems: single-phase, three-phase and stored energy. The two former are frequently called direct energy systems. Ignitron control is also used in three-phase machines in which current is drawn from all three phases of the supply instead of one. Two ignitrons and control equipment are used on each phase of the three-phase input to the transformer. This type of transformer is particularly suitable for large welding machines as it presents a more balanced load to the supply main and has a better power factor. Three-phase power sources are of two types—frequency converter or rectifier. In the former type the transformer has three primary windings, one for each phase each fed through two ignitrons in back-to-back arrangement, shown in fig. 59. One ignitron in each line passes a positive half cycle from each phase to the primary windings producing three unidirectional pulses in the secondary. The leading phase has now completed one cycle and the control current may be set to either leave the same three ignitrons on circuit, in which case another three positive half cycles will pass, or it may switch out the first bank of three ignitrons and bring in the second of each pair. If the latter is done three negative half cycles will pass and the current in the secondary will reverse. Reversal is not instantaneous as the second set of ignitrons will only begin to fire at the beginning of the next negative cycle of the leading phase. There must therefore be a delay of half a cycle at least before reversal can take place. Each complete set of

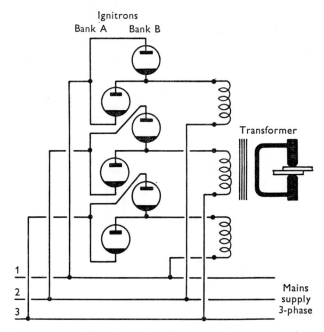

Fig. 59. Basic power circuit for a three-phase frequency changer type resistance welder.

Fig. 60. Current wave-form for a frequency changer welder.

cycles in each direction, usually either 1, 2 or 3 giving 3, 6 or 9 pulses in the secondary is called a 'heat' cycle, and the off period between reversals is a 'cool' cycle. Figure 60 shows the way the current in the secondary comprises a series of current pulses which build up exponentially because of the transformer, and also the effective change in frequency of the three-phase 50-cycle input to the secondary output. With one full cycle, three pulses, in each direction and two half-cycle cool times the highest frequency of the secondary must be $\frac{50}{3} = 16\frac{2}{3}$ c/s. Permutations of the heat times from 2 or 3 c with cool times of $\frac{1}{2}$, $1\frac{1}{2}$, $2\frac{1}{2}$ or $3\frac{1}{2}$ c allow the frequency to be

reduced down to 5 or 6 c/s. An improvement in the power factor is associated with the reduction in secondary frequency. As indicated in the next
section, a reduction in frequency lowers the reactance and, therefore, increases the relative importance of electrode/workpiece/electrode resistance.
Current control is exercised in the same way as for single-phase equipment,
there being one control in each of the three input lines.

In rectifier three-phase machines (fig. 61) the three-phase input is trans-
formed to low voltage and rectified. The welding current is therefore unidirectional d.c. but has a heavy three-phase ripple because of the lack of
smoothing and the use of phase shift current control.

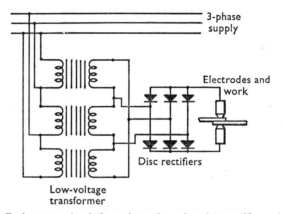

Fig. 61. Basic power circuit for a three-phase dry plate rectifier welder.

Prior to the introduction of the three-phase machine stored energy
machines were commonly used to reduce mains demand. Two types were
mainly used—electro-magnetic and electrostatic (or capacitor discharge)
according to whether the energy was stored in the transformer winding
or by charging capacitors. In both types, illustrated diagrammatically in
figs. 62 and 63, the three-phase input was rectified and on the discharge of
the energy into the primary of the welding transformer: by interrupting
the primary current with the former and discharging the capacitors with the
latter, a single pulse of energy rising sharply and decaying exponentially
was released in the secondary. With this type of machine, current and time
have a different meaning in practical terms from the a.c. case because the
energy for welding is given by the area under the single curve. Note also
that the shape of the current–time curve is as important as the area it
encloses. The shape is influenced by primary current and turns ratio for
the electro-magnetic and by capacitor voltage, capacity and turns ratio
for the capacitor discharge type.

The electrostatic type of welding machine has reappeared recently in the
form of the capacitor discharge machine for welding small components at

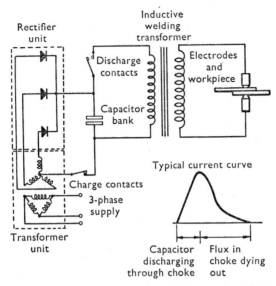

Fig. 62. Basic power circuit and wave-shape for a capacitor-discharge welder.

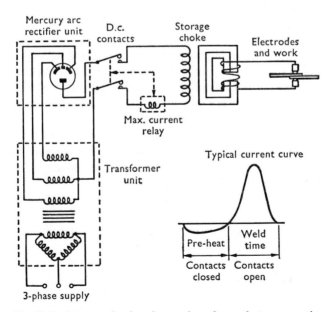

Fig. 63. Basic power circuit and wave-shape for an electro-magnetic
stored-energy welder.

short times. Welding energy is determined by the charge on the capacitor and is measured in terms of watt-seconds, not current and time separately. Such machines are used for welding dissimilar metals or delicate components where minimum fusion is required.

Factors influencing welding current. Mention has been made already of the part played by contact resistance in determining the heat generated and, therefore, weld size and strength. Even assuming correct maintenance of electrodes and consistent resistance, mains voltage and operation of controls, however, the welding current can still vary from that required, thereby causing weld size also to vary. A distinction must be made between the total current flowing between the electrodes and that actually utilized to produce the weld nugget. The welding current may be reduced because: (*a*) part of the secondary current has been diverted through a path other than that through the weld nugget, or (*b*) the secondary current itself has dropped as a result of a change in the reactance of the secondary circuit. Secondary current is dependent on the impedance of the machine circuit which is related to resistance (R) and reactance (X) thus: $Z^2 = R^2 + X^2$. Reactance is the product of inductance and frequency in the formula $X = 2\pi f L$, where f is frequency and L is inductance in henries.

Current diversion, generally called shunting, occurs when two welds are placed close together or when the design of the joint allows an alternative path between the electrodes. As a result the first weld in a line of welds may be larger than subsequent welds and the single-spot test coupon can give a larger weld than would be obtained in a line of welds in the actual job. The effects of shunting become more important with low resistivity metals and in stitch, series spot and seam welding. Shunting can be limited by adjusting the current according to the weld pitch and material resistivity or by ensuring that there is a minimum distance between spots. For normal purposes the pitch should not be less than 3*d*.

Reduction of the welding current can occur because of changes in the inductance of the secondary circuit. This factor is influenced by the geometry of the secondary or welding loop and the permeability of the material it surrounds. An increase in the secondary loop, such as might be necessary to encompass a large workpiece, will raise the inductance and therefore the impedance of the circuit so that for the same size of weld a higher primary current and kVA is required. The size of the secondary loop is normally defined in terms of throat depth.

The insertion into the throat of a mass of metal of high permeability, such as iron, absorbs energy from the magnetic field set up by the conductors and increases the inductance. A reduction follows in the current circulating in the conductors. In welding large ferrous metal components the current delivered by the machine can vary as the component is moved in or out of the throat. The extent of the current drop is not amenable to calculation but should be checked by experiment and measurement.

Except for three-phase frequency converter machines frequency is not normally a variable in resistance welding, but it may be noted that as the frequency is reduced the effects of inductance become less important. A reduction in kVA demand follows because the secondary impedance is reduced.

Timing. It is universal practice to reckon times in resistance welding in terms of cycles of the mains supply. Time is not as critical a variable as current when welding mild steel. As welding times are reduced below 10 c, however, when thin or high thermal conductivity materials are welded, increased attention to the timing device is required. Greater mechanical precision is also required (see following section on Force).

Timers can be mechanical or electronic, synchronous or non-synchronous. The latter work without regard to the phasing of the welding current and because an exact number of cycles cannot be guaranteed they are not used for times less than 10 c. Synchronous electronic timers of the digital type can give reproducible times down to $\frac{1}{4}$ c and are in wide use. Electronic timers based on resistance-capacitor circuits are also in wide use but are less reliable than the digital-type counter which is replacing them.

Another aspect of time in resistance welding is that of the correct phasing of force and current cycles in a complicated programme or the timing of a sequence of current cycles at different levels for pre-heating or post-heating.

Application of force. Providing a minimum value is exceeded electrode clamping force is not particularly critical. An average value for mild steel, the most tolerant of metals for welding, is upwards of about 10 000 lb/in.^{2} of electrode area. Materials of high strength, and particularly higher strengths at elevated temperatures, require electrode forces up to several times those necessary for mild steel. Unfortunately it is hard to generalize on the force required as part of it may be taken up in pressing the parts together or even in moving the head of the machine.

The mechanical design of the welding machine can be as important as the value of the force used. Rigidity of the frame, low inertia and lack of friction in moving parts are particularly important points. For most welding of mild steel the pneumatically operated piston with a head in simple slides suffices. As welding times are shortened, however, the mechanical precision of the whole welding machine must be improved in proportion. Welding heads may then be mounted on slides in ball or roller bearings. Diaphragms may replace pistons to give greater freedom from friction. With miniature machines leaf springs are sometimes used. All points related to mechanical design mentioned above become more critical, and consistent operation less likely in badly maintained machines or those used below their normal working range.

In a special form of resistance spot welding known as magnetic-force welding the force is applied electro-magnetically using the welding current. This can give extremely rapid application of force suitable for welding high conductivity and other metals difficult to weld by normal techniques. The equipment for magnetic force spot welding comprises the normal force application cylinder and transformer, but the secondary lead passes over a horizontal electro-magnetic core. Initial pressure is applied through the air cylinder but the piston and electrode assembly also carries the armature for the electro-magnet. Immediately before the passage of the current the magnet armature is close to the pole pieces of the core so that when current flows in the welding circuit and also energizes the magnet, additional force is applied to the electrode.

Metals which have a high shrinkage on solidification or which are hot-short, for example some aluminium alloys, may give porous or cracked weld nuggets. If adjustment of current, force and time cannot rectify the trouble a dual pressure cycle may be employed. After the current cycle and while the nugget solidifies, the force is increased to forge the nugget solid. Whenever a programmed weld cycle rather than a simple weld cycle is used the correct phasing of the force and current becomes critical. The premature application of the forge would cause excessive indentation while its delayed application would fail to cure the cracking. To assist in the forging process the weld cycle is often followed by a post-heat current cycle of lower magnitude than the welding current which delays the cooling of the weld so that it remains plastic longer.

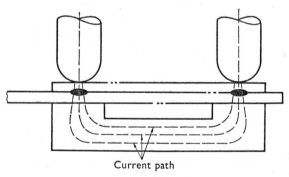

Current path

Fig. 64. Series welding.

Series welding

When it is convenient to approach the work from one side only with the electrodes or where numbers of spots have to be made at the same time a series welding technique is employed, illustrated in fig. 64. Series welding is widely used with multiple-headed machines in the automobile industry. Indirect welding is a special form of series welding in which current may enter through an electrode but leave the work through a contact pad at which point there is no weld because of the large area of contact. The effect of current shunting is of importance in series welding. There are three paths which the current can take; through the upper sheet, the lower sheet or the backing electrode. Clearly, the extent of shunting depends on the electrode spacing and can reach appreciable proportions as will be seen from table 7. The current in the upper sheet is high for the first three cycles because the interfacial resistance which forms part of the circuit with the lower sheet is initially high. At the end of the weld the current is shared almost equally between the sheets.

Table 7. *Current passing through shunt paths in series welding*

Electrode spacing (in.)	Period during weld	Upper sheet (amp)	Lower sheet (amp)	Backing (amp)
2	Start	3 200	2 000	5 500
	End	2 200	2 100	6 400
4	Start	2 300	1 600	6 800
	End	1 750	1 700	7 250
6	Start	1 800	1 450	7 450
	End	1 450	1 500	7 750

Material, 0·036 in. degreased and pickled low carbon mild steel, total welding current 12 c at 10 700 amp, 6 in. radius electrode, electrode force 380 lb. (Data from 'Measurement of shunting currents in series spot welding 0·036 in. steel', by E. F. Nippes, W. E. Savage and S. M. Robelotto, *Weld. Res.* (Suppl.), **20**, no. 12, 1955, pp. 618–24.)

Heat balance

There are frequently occasions when metal of two different thicknesses or compositions must be joined. Such differences result in greater heat generation or abstraction on one side than the other and the nugget may grow with its centre-line away from the interface resulting in a weak weld. In joints with sheets of equal thickness but unequal resistivity and conductivity the nugget will grow towards the high resistivity side. Where similar materials are welded, but the thicknesses are unequal, the nugget grows towards the thicker side. Equality of fusion into the two different materials may be achieved by increasing the heat generation in the thick or high conductivity metal. This is done by using a smaller diameter electrode (to increase current concentration) or one with a high resistance insert (to reduce heat loss) in contact with the thick or the high conductivity metal. If the thick metal is also the one with the higher conductivity the effects of thickness and conductivity can be compensating. With sheets of equal thickness, the nugget would normally move into the low conductivity sheet where most heat is developed. Increasing the thickness of the metal with the higher thermal conductivity moves the nugget in the direction of that metal because the total heat loss along that path is reduced.

Types of equipment

Rocker arm and press type. The simplest arrangement for resistance spot welding is one in which a single pair of electrodes is used and the work is offered to the machine by hand. Equipment for this type of welding is of two types: (1) rocker-arm welder, (2) press welder. The diagram of the rocker-arm type in fig. 65 gives the salient features. Although the sketch

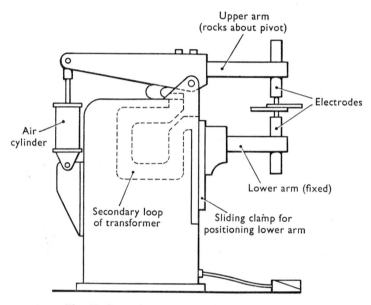

Fig. 65. General arrangement of rocker-arm welder.

shows an air-operated arm both foot- and cam-operated machines are used. With the latter types the force must be applied through a spring. Electrode force depends on the mechanical advantage offered by the lever ratio and the setting of the spring. Rocker-arm machines are low cost, simple machines suitable for welding light-gauge mild steel with simple welding cycles. High electrode forces cannot be transmitted with this type of equipment, however, so that for welding thick material or alloys requiring precise control, the press-type machine shown in fig. 66 is preferred. Press welders are also extensively used with mechanized work feeds. Because the electrodes in the rocker-arm welder do not close axially scuffing of the electrode surface can occur and electrode wear is higher than with the press welder. A variety of different electrode shapes is used with both types of machine to give access to the joint. Although frequently essential, offset electrodes wear more quickly and are more difficult to maintain.

Multiple welding. Multiple welding techniques employ a number of electrode assemblies grouped in such a way that a whole assembly may be welded in one operation. The type of multiple welder now mainly in use has each electrode assembly connected to a separate transformer through flexible leads, both transformers and electrodes being mounted in machine tools which allow a complete component to be handled. Many such welding machines resemble large press tools with the electrodes mounted on the platens and the electrodes are hydraulically operated and served from a common system. Series welding is often used and then two or four electrodes may be worked from one transformer. Because the machines are highly specialized the electrical and hydraulic systems can be

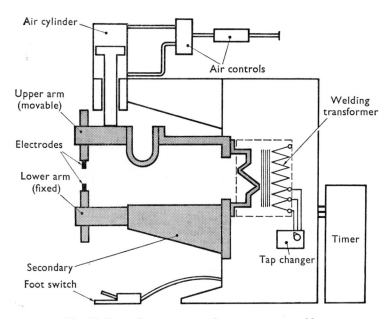

Fig. 66. General arrangement of press-type spot welder.

designed specifically for welding a limited range of metal thicknesses. Multiple welding is widely used in the automobile industry.

Gun and portable welders. There is frequently a requirement in the assembly of large components by welding for a portable welder which can be brought to the job. Because of the wide variety of applications there are many different designs of portable welder but these are of well-defined types. The welding head or gun in all but the smallest units is separated from the transformer by a flexible cable. Welding guns are of two types according to the method of clamping the electrodes. Many guns, particularly the lighter ones, are of the pincer type using levers as in the rocker-arm machines and the other type is the C clamp gun in which the force is applied directly through a piston, the portable equivalent of the press welder. Although the smaller type of pincer gun may be manually clamped, larger guns are slung from balancing booms so that the operator is freed from the strain of directly supporting the gun. Methods of supporting the gun and transformer to give the maximum degree of freedom is illustrated diagrammatically in fig. 67.

The requirement of portability and flexibility demands particular attention to welding cables. Circuit reactance is kept to a minimum in spite of the long cables by running an equal number of conductors in and out, in

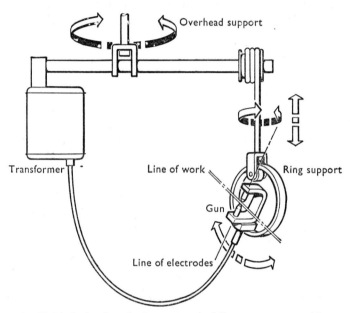

Fig. 67. Methods of conferring manoeuvrability on gun spot welders.

a symmetrical arrangement down a single cable. The close spacing ensures minimum reactance as well as preventing kicking of the cable which would otherwise occur when the current passes. For this reason such cables are often called 'kickless' cables, although their primary function is to be of low reactance. Water cooling of conductors enables their size to be reduced. Inevitably the secondary resistance in a gun welder is higher than for a press-type welder so that higher secondary voltage transformers are required. Current control is usually by simple primary tap switching as the transformers are single phase and the work range is limited.

Applications

It might be thought that the spot- and seam-welding processes have a restricted field of application because of the limited variation in joint design which is permissible—that is, lap joints in sheets of the same order of thickness. The processes, however, have extensive application in the joining of sheet metal not only in mild steel but also in stainless steels, heat-resisting alloys, aluminium and copper alloys and reactive metals. Dissimilar metal combinations are also welded. Seam welding is not often carried out on metal thicker than ⅛ in. because of the mechanical difficulties in applying pressure and wheel wear, but spot welding has been used on steel as thick as ¾ in. Such applications are rare, however, requiring massive machines and long weld times so that the normal upper limit is in the region of ¼ in. Particularly attractive features of the resistance spot-welding process are the high speed of operation, ease of mechanization, the self-jigging nature of the lap joint and the absence of edge preparation or filler metal.

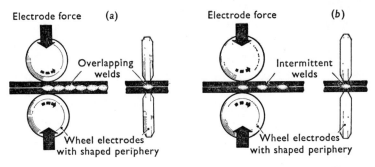

Fig. 68. Seam-welding principles: (*a*) overlapping spots; (*b*) roller spot.

RESISTANCE SEAM WELDING

When a continuous seam is required, as opposed to an array of individual spot welds, two methods are available. Using the equipment described already a series of overlapping spots can be made—a process known as stitch welding. Alternatively, the electrodes may be replaced by wheels or rollers so that work may be moved through the welder continuously without the necessity for raising and lowering the head between welds. The rollers are power driven and may or may not be stopped while individual welds are made (fig. 68). Current is generally passed intermittently while the electrodes are stationary, partly because trouble with current conduction through moving parts is reduced, but continuous current is also used to a limited extent. A common technique is known as step-by-step seam welding, because while each weld is being made rotation is stopped and the current is then switched off while the rollers move to the next position. The amount of overlap between spots depends on peripheral roller speed and the ratio of current on-and-off time. A normal overlap would be 25–50 per cent.

Adjustment of timing can be made to produce not a continuous seam but a series of individual welds. When this is done the process is called roller-spot welding. It will be seen, therefore, that spot and seam welding are very similar and that the terminology refers to the resultant weld. The real distinction is between the use of spot electrodes and roller electrodes.

Weld size depends on contact area and hence on the track width of the rollers and their diameter. Rollers are bevelled in the same way and for the same reasons as spot electrodes. Because they cannot be cooled internally roller electrodes may be flooded with water externally. Current is lost through the shunt path provided by the previous welds so that relatively higher currents are required for seam welding than for spot welding. The high duty cycle and continuous operation of a seam welder requires that the electrical equipment should be rated accordingly.

Special forms of seam welding

Generally seam welds as well as spot welds are made on overlapped sheets and the final product is a lap joint. Often, however, a butt joint is required. The fact that current and force are applied normally to the work surface and joint interface would preclude this, but special techniques can be used to overcome this difficulty.

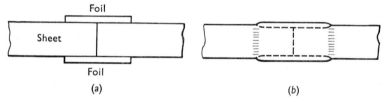

Fig. 69. Diagrammatic cross-section of a foil butt-seam weld: (*a*) before, and (*b*) after welding. Nugget forms between hatched lines.

Mash-seam welding. If the overlap in seam welding is reduced to no more than $1\frac{1}{2}$ times the material thickness the overlap can be forged or 'mashed' down while passing through the rollers. As distinct from normal seam welding the current is reduced so that excessive melting is avoided, the object being to produce a suitably hot area for the rollers to forge down. Considerable accuracy is required in presenting the work which must be restrained by clamps or rollers from sideways motion. Mash-seam welding is used mainly for the manufacture of barrels and domestic equipment in mild steel where straight welds are possible and appearances demand a flush weld.

Foil butt-seam welding. Similar applications but also including the welding of panels are possible with foil butt welding. In this process narrow foil 0·008 in. thick is fed between the seam-welding rollers and the joint in the work which is close butted (fig. 69). Although the joint interface is parallel to the direction of current flow heat is generated at the two interfaces between foil and work which are normal to the current flow. As a result both workpieces and foil are fused together to give a weld which is only slightly thicker than the parent metal. Accuracy of presentation is necessary, but less lateral support is required than with mash welding because of the more modest forging. The simpler and less rugged jigging allows large panels to be welded by this process compared with the more restricted range of the mash-seam weld process.

Resistance butt-seam welding. Large quantities of steel tube are made by the E.R.W. method, a resistance butt-seam welding process, from strip which is continuously edge sheared and rolled into tube form to present a longitudinal seam. The welding head takes the general form indicated in

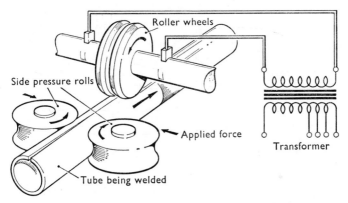

Fig. 70. Electric resistance butt-seam welding of tube from strip (E.R.W. process).

fig. 70. Current of up to 40000 A at about 5 V is introduced across the joint by the split electrode rollers and force applied by the pressure rolls. A rotating transformer with slip rings on the primary side can be used to overcome the difficulty of introducing the heavy welding currents directly to the moving electrodes. This application is seam welding in which both work motion and current are continuous. The maximum speed of the mill is limited by the welding current frequency since, as the welding speed is increased, the individual half cycles of current would eventually produce a series of disconnected spots instead of a continuous weld. Frequencies up to 350 c/s are used in practice with speeds of 120 ft/min. The completed tube has a fin of extruded metal along the weld line both inside and out which is removed continuously by cutters.

High-frequency resistance welding. In butt-seam welding, just described, heat is generated mainly by interfacial contact resistance as in spot welding or projection welding. By increasing the frequency of current supply very considerably to about 450 kc/s and raising the voltage from units to tens, a superficially similar, but in fact rather different, process was developed.

Whereas the strip is formed into tube in a mill similar to that for low-frequency welding, the welding current is introduced not through a split roller but through two probes which make light contact on either side of the joint. At the frequency used the skin effect by which the current flow tends to concentrate at the surface of the conductor becomes marked. The depth of the layer in which most of the current flows is proportional to $\sqrt{(1/f)}$ for any given material. Contact with probes is made a short distance before the two sides of the joint are forged together as indicated diagrammatically in fig. 71. Because of the skin effect the current path between the probes lies along the edges of the strip through the apex of the V formed by the edges as they close. The depth of the heated region is extremely

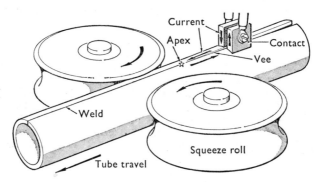

Fig. 71. High-frequency resistance butt welding of tube.

shallow, generally less than 0·030 in. and it is in precisely the best position for welding. As the joint closes the heated edges are forged together to give a high quality weld. Superficial melting can occur and this thin molten layer is squeezed out as the edges meet. For this reason the high-frequency resistance process is capable of welding non-ferrous metals and others which form refractory oxide skins. With the low-frequency process melting does not occur so that considerable deformation would be required to rupture the oxide films and give a good pressure weld. It is also difficult to achieve a sufficient temperature gradient with a high conductivity metal. In the high-frequency process surface films are flushed out with the molten metal.

Because of the high voltage (about 100 V) and the high frequency at which the current is supplied there is no difficulty in achieving good electrical contact with the probes, even on scaled material. The water-cooled probes can weld tens of thousands of feet of tube before being replaced for wear. Another consequence of the high voltage, a result of the long current path, is that high power levels can be obtained with relatively low currents. The working range, depending on material thickness and speed, would be 200–5000 A. Using a 60 kW power unit, tube of material 0·025 in. thick can be welded at up to 300 ft/min. Welding speed depends on tube thickness and not on diameter.

Although the main use of the process is for the continuous welding of tube it is clear that the principle of the method is important and has a much wider potential. Lap, corner and T welds can be made; in fact, any type of joint in which the requirements indicated in fig. 71 can be provided.

High-frequency induction welding. This method also has been used for welding tube and resembles the high-frequency resistance process in that use is made of the skin effect. The difference, however, is that instead of direct contact being made with the work the current is induced in the surface layer by a coil wrapped round the formed tube. Surface heating and fusion occur and the weld is consolidated by a forging action on the joint. Induction welding is not limited to tubes and may be applied to other symmetrical assemblies in which the joint forms a complete loop, for example as in the welding of a cap to a tube. With this type of joint there is no forging, the edges of the component merely being allowed to melt and run together. The process is not suitable for welding high-conductivity metals or those with refractory oxides as there is no active mechanism for oxide dispersal. High-frequency induction heating-type equipment is used and fusion is completed in a few cycles of mains frequency.

PROJECTION WELDING

Current concentration is achieved in this process by shaping the workpiece so that when the two halves are brought together in the welding machine current flows through limited points of contact. With lap joints in sheet a projection is raised in one sheet through which the current flows to cause local heating and collapse of the projection. Both the projection and the metal on the other side of the joint with which it makes contact are fused so that a localized weld is formed. The process is illustrated in fig. 72. Because current concentration is carried out by the workpiece the shaped electrodes used in spot welding can be replaced by flat-surfaced platens. These not only conduct the current to the workpiece they also give support so that there is no deflection except at the projection.

Projection welding is not limited to sheet–sheet joints and any two mild steel surfaces which can be brought together to give line or point contact can be projection welded. Projections can be artificial—produced deliberately by pressing and machining (see fig. 72) or they can be formed by the natural contours of the parts to be joined (fig. 73). Excellent examples of the latter are the welds between crossed wires in wire mesh or between sheet and rod as in the welding of the wards of a key to its shaft. Unlike spot welding there is the possibility of making not only lap welds, but many other types of joint as well. The welds do not have to be individual spots but may be elongated or even annular.

Similar equipment is used to that for spot welding except that the cylindrical electrodes are replaced by flat copper platens or dies. These may have inserts of wear-resistant high-conductivity copper alloys to increase the working life of the die surfaces. It is common for three projection welds to be made at one time and even groups of four or five have been attempted. With more than three simultaneous welds, however, there is a tendency for lack of consistency as, unless special pressure-equalizing dies are used, it

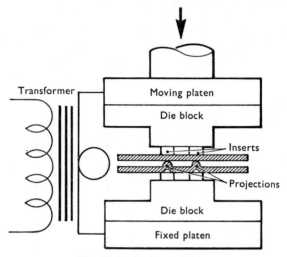

Fig. 72. Projection welding.

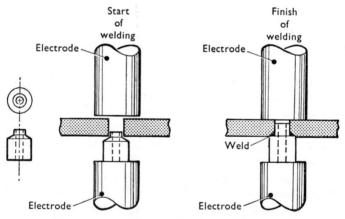

Fig. 73. Annular projection welding of hollow stud to plate.

cannot be guaranteed that all the projections will behave identically. Because several welds can be made simultaneously there is no shunting problem, as in spot welding. The method is therefore satisfactory for designs where several welds must be made close together.

Behaviour of projections

The history of projection collapse from the moment of initial contact with the other part of the workpiece is crucial to the production of a weld. There is still much to be learnt about this most complicated aspect of the process, although much experience has permitted the design of projections which give satisfactory welds. With spot welding the important process variables controlling weld size are electrode tip diameter, current, time, force, electrical resistivity and thermal conductivity of metal and surface resistance. The electrode diameter does not apply in projection welding

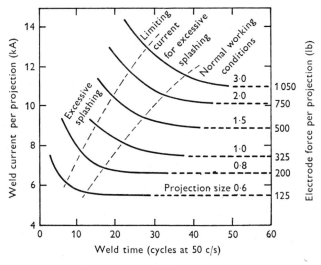

Fig. 74. Current–time relationship for projection welds in mild steel sheet. Projection sizes 0·6–3·0 used in sheet from 0·021 to 0·13 in. thick.

and surface resistance is of reduced importance, but we may add to the list of variables, projection diameter, height and shape and the strength/temperature properties of the metal being welded. Figure 74 summarizes welding conditions for a range of mild steel sheet.

When the initial load is applied the projection should not suffer excessive 'cold collapse' as this will reduce the current concentration and hence the heat generated during welding. Low electrode forces are, therefore, preferred and additionally the velocity of approach of the welding dies should be controlled to avoid damaging projections by impact.

On passing the current the projection heats up, the metal softens and collapse takes place often in less than one cycle of current. The welding current and the rate of follow up of the welding head must be such that during this stage the hot metal bridge does not become molten and splash. On the other hand, excessively rapid collapse of a projection will limit the heat generated. Projection height, diameter and shape must therefore vary systematically with metal thickness. The process does not end with the collapse of the projection, however, this stage is of critical importance because the hot slug of metal formed during collapse acts as a marker for the nugget. As with spot welding the temperature coefficient of resistance ensures that the initial hot spot is the seat of development of the final nugget. Plate 10 summarizes the process with oscillographic records.

The requirement for movement of the welding head to follow projection collapse is a major distinction between spot and projection welding. Not only does such motion, termed 'set-down', require that the component design and projection spacing allow movement but it also demands greater attention to machine performance. Thus, although spot and projection welders are superficially similar the latter require more attention to lack of friction and minimum inertia in moving parts. It is also clear that until set-down has occurred there is no pressure seal around the weld, as in spot welding, to retain molten metal.

With so many conditions to be satisfied simultaneously projection welding in practice is almost confined to applications in mild steel, a metal especially suitable for forging and tolerant to welding conditions. Except for the soft high-conductivity metals, in which projections of adequate strength are difficult to produce, most metals should be amenable to projection welding if suitable control could be exerted over the process. One method of control which has been found effective, with cross-wire welding of aluminium for example, is current slope control. In this technique the current is increased gradually from the beginning of the weld time over several cycles to the welding value. The current–time characteristic of the projection collapse is now not wholly dependent on the collapse mechanism itself. Many micro-welding applications involve a form of projection welding, often crossed wire, and here capacitor discharge machines can be an advantage because the current build-up depends on the capacitance and impedance of the circuit and can give an inherent slope control.

Heat balance

Because heat is generated during projection collapse there is a tendency for the part of the workpiece containing the projection to become hotter than the other. Where unequal thicknesses are to be joined, therefore, the projection is placed in the thicker material. Similarly, if different composition materials are being welded together heat balance is improved by placing the projection in the metal with the higher thermal conductivity. As with spot welding, if one electrode has a low thermal conductivity the nugget will move toward that electrode. These are the main methods by which heat balance is controlled. See also heat balance in spot welding.

Applications

Projection welds are used singly or in small groups unlike spot and seam welds in which an array of many welds may be used to make a joint. Only with annular projections is there any attempt to make a seam, but this must be made as a single event and the seam is unlikely to be longer than 10 in.

The process is used extensively for making attachments to sheet and pressings and these attachments may be either sheet metal or solid parts. Joints of the latter type could not be made by normal spot welding. Other important applications are the joining of small solid components to forgings or machined parts. Crossed-wire projection welding is widely used. A great variety of joints is possible, the limitations being chiefly those of material and the ingenuity of the designer to devise projections to satisfy the conditions for heat generation. Because there is no possibility of using post-weld current pulses for heat treatment weld assemblies in hardenable materials must be heat treated in a furnace. Apart from mild steels projection welds may be made readily in alloy steels and titanium alloys.

FLASH WELDING

The flash-welding process was developed from resistance butt welding probably by accident in attempts to increase the capacity of the butt welding machines by raising the voltage and applying pressure intermittently. Similar equipment is used for both processes. This comprises one fixed and one movable clamp, so that the workpieces may be gripped and forced together, a heavy-duty single-phase a.c. transformer with a single-turn secondary, and finally, equipment for controlling current, movement, force and time. The parts to be joined when gripped by the clamps and brought into contact complete the secondary circuit (see fig. 75). When the welding voltage of up to about 10 is applied at the clamps current flows through the initial points of contact causing them to melt. These molten bridges are then ruptured and small short-lived arcs are formed. The platen on which the movable clamp is mounted is moving forward while this takes place and fresh contacts are then made elsewhere so that the cycle of events can be repeated. This intermittent process, during which much of the metal contained in the molten bridges is expelled violently in a spectacular manner, is called 'flashing'. Flashing is allowed to continue until the surfaces to be joined are uniformly heated or molten. By this time the moving platen will have advanced, at an increasing rate, to close the gap as metal is expelled, the total distance up to the point of upset being known as the flashing allowance. At this point the rate of movement of the platen is rapidly increased and a high force applied to forge the parts together and expel the molten metal on the surfaces. The metal expelled forms a ragged fin or flash round the joint. Ideally, all the molten con-

taminated metal produced during flashing should be removed in this way
to produce a high-quality joint having many features of a solid-phase weld.

Flashing

The conduct of the flashing process is of extreme importance in achieving
satisfactory welds. Flashing cannot begin until a molten bridge has been
formed somewhere along the interface. If the parts fit with accuracy difficulty in initiating the process may be experienced and it may be necessary
to bevel the butting surfaces to give local contact. Note that this procedure
is very different from that for resistance pressure welding where a good fit

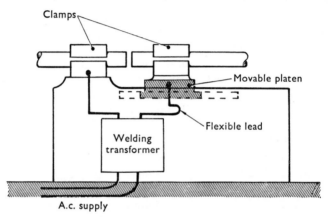

Fig. 75. Basic arrangement for flash welding.

is desirable to exclude the atmosphere and prevent surface contamination.
In flash welding surface contamination is removed in the spatter during
flashing and molten metal is expelled in the final upset or forging operation.

Once a molten bridge has been formed it is set in motion over the interface by the magneto-dynamic forces resulting from the distortion of the
current path through the workpieces. While it races over the interface at
speeds of 10 cm/s or more heat is generated in the bridge by the joule effect
and is lost to the colder metal at each end. Assuming that the bridge
does not freeze, it will be broken in one of two ways. It may either explode
because of the combined influence of overheating and the pinch effect or it
may race to the edge of the workpiece where it becomes extended in the
form of an arch between the work faces until it ruptures. In either event
much of the metal contained in the bridge is propelled out of the joint as
spatter. Because there is inevitably considerable inductance in the secondary circuit the collapse of the electric field at rupture results in a voltage
surge which is capable of initiating an arc, although the open circuit voltage cannot sustain it and it is rapidly extinguished. Considerable heat is
generated in these brief arcing periods which accounts for the efficiency of

the process compared with resistance butt welding. Oscillographic records of welding conditions indicate that the active periods in the process tend to occur at the periods of peak supply voltage and that there are frequently long periods of inactivity.

Process variables

With a picture of the flashing process the influence of the process variables can be appreciated. The instantaneous flashing speed will be seen as a critical variable. If this is too high the molten bridge will be quenched prematurely, resulting in a tendency for the workpieces to freeze, but if the rate is too low the periods of inactivity will be increased. At any point in the flashing cycle there is an optimum rate at which the movable platen must be advanced and this rate increases progressively throughout the cycle as the temperature builds up in the workpieces. This increase in rate of platen advance is achieved by the use of a cam or similar device in mechanical machines or by a variable flow valve in hydraulic equipment. Within the working limits an increase in flashing speed, obtained by using higher platen speeds, results in a higher current demand and a greater heat input rate. The system is analogous to a consumable electrode arc system, and the flashing process in particular has some features in common with the short-circuit arc processes.

A minimum voltage is required to sustain the flashing process which decreases slightly as the surface temperature of the workpieces is raised. Since current control is by the setting of the turns ratio on the transformer through a primary tap switch, high voltages are associated with high currents. Setting the current too high, therefore, results in greater ease of maintaining the micro-arcs formed in the flashing period and deep craters can be caused in the flashing face. The electro-magnetic forces controlling metal expulsion, dependent on the square of the current, are also increased with high welding currents. Deep empty craters formed in the latter stages of flashing may not be filled with molten metal or closed during forging so that dangerous regions of lack of fusion, called 'flat spots', may be left in the plane of the weld. On large machines when difficulty may be experienced in getting the flashing process going the equipment may provide for a high voltage initially which is reduced once the process is proceeding.

The time taken over the flashing process directly affects the loss of metal and, therefore, the flashing allowance. Of greater importance in relation to quality, however, is the effect of flashing time on surface temperature and temperature gradient. The surfaces must be reasonably uniformly heated and if not actually molten all over then at least the metal should be highly plastic and close to the melting point. Away from the flashing surface the temperature gradient should be such that when the upset force is applied the molten metal can be expelled and flow can take place at the interface. Short times and high flashing speeds give steep temperature gradients and long times and low flashing speeds allow heat to spread. The narrow upset band associated with steep temperature gradients may not allow efficient expulsion and the abrupt change in direction of structure and inclusions can give poor mechanical properties. Apart from the wastage of metal long times are to be avoided because too great a spread of heat may result in a general swelling of the joint region during upsetting and inhibit the expulsion of metal, as indicated in fig. 76. The

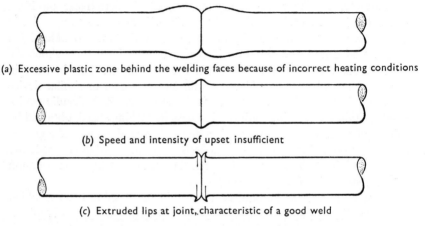

(a) Excessive plastic zone behind the welding faces because of incorrect heating conditions

(b) Speed and intensity of upset insufficient

(c) Extruded lips at joint, characteristic of a good weld

Fig. 76. Influence of process variables on the condition of the upset metal in flash welding.

temperature gradient can be influenced by the proximity of the clamps and the thermal conductivity of the metals being joined. With metals such as aluminium or copper alloys welding currents tend to be higher and times shorter than for mild steel, and conversely for metals with lower thermal conductivity and higher hot strength such as stainless steels.

The flashing process with its associated wastage of metal may be shortened with advantage, particularly for sections over 0·5 in. cross-section, by preheating the workpieces. This may be achieved by bringing the parts into contact a number of times for periods of about 1 or 2 s. Between the periods of contact the workpieces are withdrawn so that heating is by electric resistance only and flashing is delayed until the temperature has built up.

Heat balance

So that similar temperatures may be reached and plastic flow can take place on both sides of the joint line there must be a heat balance. If a part with a small cross-section must be joined to one considerably larger, the larger part must be prepared so that the weld is made between similar

Table 8. *Flash welding parameters for mild steel
extended sections (no preheating)*

Thickness (in.)	Flashing allowance (in.)	Upset allowance (in.)	Flashing time (s)
0·125	0·50	0·19	8
0·250	0·75	0·28	18
0·500	1·00	0·40	46
1·000	1·25	0·50	110

areas. Frequently, dissimilar metals are joined by flash welding and allowance must be made for differences in thermal conductivity. If two such parts are disposed symmetrically between the clamps the weld would tend to finish up closer to the clamp holding the low thermal conductivity metal. By arranging for the low conductivity metal to protrude further from its clamp flashing can be continued until the high conductivity part reaches an adequate temperature.

Upsetting and clamping

The upsetting operation follows immediately on flashing, the current not being cut off until upsetting commences. This is required so that the temperature is maintained and adequate plastic deformation can take place without cracking during upsetting. Heating during this period means that lower upset forces are required. Upset forces depend in addition upon flashing speed since a high speed gives narrow heating and necessitates higher forces to give plastic flow. Strength/temperature properties of the materials welded are directly related to upset forces, the average figure of 10000 lb/in.^{2} for mild steel being increased by up to four times for high hot-strength materials. The upsetting process causes a decrease in length, the upset allowance, which can be from 25 to 40 per cent of the flashing allowance, the higher figure being more usual with smaller cross-sections. Since it is the transverse plastic flow induced by the force which is the significant factor the upset allowance could be the more meaningful parameter.

Similar considerations as regards welding machine rigidity and the free movement and inertia of moving parts apply for flash welding as for resistance spot welding. With flash welding the workpieces must be held by the clamps with enough force to resist slip. This requires a clamping force of up to twice the upset force. Where possible, as with short pieces for example, arrangements are made for the workpieces to be held against a seating or back-up which reduces the clamping requirements (pl. 11). Clamps must also provide suitable electrical contact for the heavy welding currents and usually have wear-resistant copper alloy inserts in water-cooled blocks. Adequate area of contact and cleanliness of the stock are necessary if local arcing or 'die burns' are to be avoided. The length of metal projecting beyond the clamps should be short enough to resist buckling under the upset force but long enough to minimize the chilling effect of the clamps.

Applications

Flash welding is particularly suited to joining bar stock and similar compact sections, however extended sections such as pressings or joints in strip from continuous rolling mills can also be made. Where thin sections are joined high-quality jigging is required to secure accurate alignment. Other common joints are bar stock or tube to forgings as well as welds

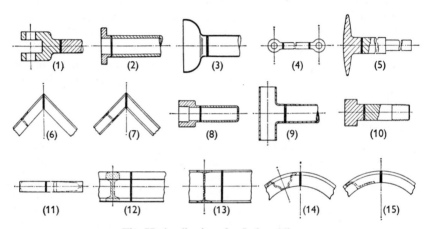

Fig. 77. Applications for flash welding:

(1) Connecting rod	(6) Angle	(11) Drill blank
(2) Flange	(7) Metal casement	(12) Track rail
(3) Housing	(8) Adapter	(13) Beam
(4) Tie bar	(9) Tube T piece	(14) Auto wheel
(5) Buffer end	(10) Bolt	(15) Heavy rim

between forgings as illustrated in fig. 77. Many types of rings and frames are joined by flash welding, e.g. wheel rims, aircraft engine rings and window frames.

This type of application calls for attention to three points of detail: (1) A ring provides a shunt path for the current for which loss an allowance must be made. (2) The motion during flashing requires that the ring is sufficiently flexible to allow movement of the clamps. (3) With rings or frames, or indeed with many other components, accuracy in maintaining the total allowance (flashing and upset) as required by the design is necessary so that the shape of the welded product can be controlled. When a ring is too small and heavy to permit conditions (1) and (2), as in chain making, it can be made by joining two halves with two welds made simultaneously.

Mild, carbon and alloy steels are extensively welded by flash welding as well as aluminium alloys and other non-ferrous metals. Carbon steels containing more than 0·4 % C, or alloy steels with equivalent hardenability require post-weld heat treatment to prevent excessive formation of the hard martensite phase in the weld and heat-affected zone. This may be achieved by delaying the cooling as a result of passing one or more pulses of current after upset has taken place. 'After removal from the clamps the parts may be transferred to a bath of powdered insulating material to delay cooling still further. The method is sometimes called post-weld annealing. Alternatively, some steels can be allowed to cool low enough for the martensite transformation to take place, this structure then being tempered by a single post-weld current pulse.

The absence of cast metal makes the process suitable for welds between dissimilar metals and in this respect flash welding has points in common with solid-phase methods. One of the important dissimilar metal applications is that used in drill blanks between high-speed steel and carbon steel. This is also an application in which post-heating is used.

The essentially automatic nature and high speed of flash welding make it a mass production process when suitable tooling is provided. Manual application of the upset force through levers has been used for many years on small sizes, but with few exceptions the advantage of consistency in the product leads to preference for the automatic machine. While simple shapes can be welded on standard machines many special-purpose machines are produced. Examples of such equipment are the machines for welding rails, steel strip, window frames and automobile rear-axle casings.

Bibliography

Begeman, M. L. and Funk, E. J. (1955). Seam welding dissimilar thicknesses of low-carbon steel. *Weld. J.* **34**, no. 11, 529S–34S.

Bentley, K. P., Greenwood, J. A., Knowlson, P. M. and Baker, R. G. (1963). Temperature distribution in spot welds. *Br. Weld. J.* **10**, no. 12, 613–19.

Blair, J. S. (1953). The electric resistance welding process for making steel tubes. *Trans. Inst. Weld.* **16**, no. 5, 117–26.

Busse, F. (1961). Foil seam welding. *Br. Weld. J.* **8**, no. 4, 123–9.

Dawes, C. J. (1966). Quality control in resistance welding. *Assembly and Fastener Methods*, **4**, no. 3, 32–6.

Hörmann, E. (1960). High-frequency resistance welding with contact electrodes. *Schweissen und Schnieden*, **12**, no. 10, 431–8.

Humpage, R. W. and Burford, B. C. R. (1950). Formulae for obtaining optimum settings for the spot welding of clean mild steel. *Welding*, **12**, 515–22.

Knowlson, P. M. (1965). Instrumentation for resistance welding. *Br. Weld. J.* **12**, no. 4, 167–90.

Knowlson, P. M. (1966). Projection geometry and weldability. *Br. Weld. J.* **13**, no. 9, 536–57.

Kouwenhoven, W. B. and Sackett, W. T. (1950). Spreading resistance of contacts. *Weld. Res.* (Suppl.), **27**, no. 10, 512S–20S.

McDowall, K. H. and Weeks, D. A. (1959). A practical approach to resistance welding. *Br. Weld. J.* **6**, no. 9, 381–95.

Nippes, E. F., Savage, W. F., Grotke, G. and Robelotto, S. M. (1957). Studies of upset variables in the flash welding of steels. *Weld. J.* **36**, no. 4, 192S–211S.

Phipps, G. A. and Knowlson, P. M. (1964). Review of projection welding in mild steel sheet. *Br. Weld. J.* **11**, no. 1, 28–39.

Recommended practices for flash welding (1963). *Br. Weld. J.* **10**, no. 1, 10–16.

Resistance Welding Manual, 2 vols. Resist. Weld. Man. Assoc. (U.S.A.).

Rudd, W. C. (1961). New high-frequency resistance welding applications. *AIEE Paper*, pp. 61–647.

Shillam, G. D. (1963). Factors influencing the choice of spot welding in design. *Light Metals Industry*, **26**, no. 306, 44–6.

Taylor, H. G. (1955). A history of resistance welding. *Trans. Soc. Engineers (Inc.)*, **66**, no. 1, 11–25.

Waller, D. N. (1964). Head movement as a means of resistance-welding quality control. *Br. Weld. J.* **11**, no. 3, 118–22.

7

THERMO-CHEMICAL WELDING PROCESSES

Welding processes in which the parts to be joined are fused by the heat of chemical reactions are of two types—those employing flames and those in which exothermic processes such as the aluminothermic reaction are used. Both methods had their origin at the turn of the century. Welding with flames, or gas welding as it is frequently called, was launched as a result of the development of the Linde method for producing oxygen from liquid air in 1893 and the oxy-acetylene torch devised by Fouche and Picard in 1903. The exothermic reactions between aluminium powder and metal oxides demonstrated by Sainte-Claire-Deville and then studied by Vautin in 1895 were used a few years later by Goldschmidt for the reduction of ores and for welding. This method of joining is known as Thermit welding.

GAS WELDING

From the welding point of view one of the most important character-istics of a flame is its temperature as this largely determines the rate at which welding can be carried on. Heat is transferred from the flame to the work by forced convection and by radiation. The former is proportional to the flow of gas and the temperature difference between the gas and the work, and the latter, as defined by the Stefan–Boltzmann law, to the fourth power of the absolute flame temperature. The dependence on the fourth power of the temperature would indicate that a small increase in flame tem-perature should give a marked increase in welding speed. It has been found, however, that only about 15 per cent of the heat available for welding is transferred by radiation so that flame temperature is not the sole criterion by which a fuel gas is judged. A method which takes account of flame energy and flow has been proposed in what is called combustion intensity. This quantity is the product of the burning velocity or flame speed of the mixture used and its heating power and may be expressed in B.t.u./s/ft^2 of flame cone area. Combustion intensity can be calculated for total com-bustion or for the primary stage only.

Flame temperature depends largely upon the net calorific value of the fuel gas and the volume and specific heat of the products of combustion. A high net calorific value (gross c.v. less the heat of vaporization of any water formed), however, may be offset by the requirement of a large volume of oxygen to complete combustion. The volume of oxygen required to complete combustion of a unit volume of fuel gas is called the combustion ratio. This oxygen absorbs some of the heat of the reaction and therefore reduces the flame temperature. The presence in the flame of an inert gas such as nitrogen, which is also heated but takes no part in the reaction, is a further reason for a reduction of temperature. Flames are hotter, therefore, when the fuel gas is burnt in oxygen than in air. Temperatures are also reduced if an excess of either fuel gas or oxygen are used as may be required for metallurgical reasons. The combustion reactions are reversible at high temperatures so that they are not complete in a flame. This dissociation reaction means that unburnt fuel gas and oxygen in the flame also carry away theoretically usable heat. The characteristics of three important fuel gases are given in table 9.

Table 9. *Fuel gas characteristics*

Fuel gas	Net c.v. (B.t.u./ft³)	Combustion ratio	Ratio of oxygen to fuel gas used	Flame temp. (°C)	Combustion intensity (Total) (B.t.u./s/ft²)
Acetylene	1433	2·5	1 to 1	3250	12700
Propane	2309	5·0	3·25 to 1	3100	5500
Hydrogen	275	0·5	0·5 to 1	2800	7500

Note 1. Table compiled from data given by C. G. Bainbridge, 'Gas welding and cutting', *Welding Handbook*, 3rd ed., 1950, Moen and Campbell.

Note 2. The ratio of oxygen to fuel gas used in a welding flame differs from the theoretical combustion ratio because some of the oxygen required for combustion is supplied by the air in which the flame burns.

The table shows that acetylene gives the hottest flame and has the highest combustion intensity, and it is mainly for these reasons that that gas is favoured for welding as they permit higher welding speeds.

It is the function of the welding torch to bring together correct volumes of the fuel gas and oxygen, mix them efficiently and pass them through a nozzle, to form a flame with characteristics suitable for welding. Control of the gases is generally done by two valves in the handle of the torch shown in fig. 78. Mixing takes place in a chamber into which the fuel gas is passed through an orifice surrounded by several other holes through which the oxygen is passed. The aim is to produce maximum turbulence and therefore the most efficient mixing. At the nozzle, however, laminar flow is desired to give a smooth flame and this is achieved while the mixed gases pass through the torch to the tip of the nozzle.

The design of the torch nozzle is important in a number of respects. Torches are required for welding a variety of metals and different thicknesses which require varying heat inputs. The heat or 'flame power' is determined by the volume of fuel gas burnt by the torch in unit time —that is on both the pressure of gas and the diameter of the orifice in the nozzle through which it issues into the flame. Pressure and nozzle diameter also determine gas velocity and cannot be varied independently over a wide range.

If the pressure is too high for a particular size of nozzle the flame will be harsh and turbulent, possibly tending to disturb the weld pool. When the pressure is low the flame becomes soft. The gas velocity through the nozzle must always exceed the velocity of the flame in the mixture used, otherwise the flame will back-fire. Conversely, if the gas velocity at the

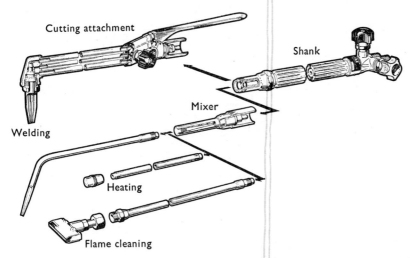

Fig. 78. Oxy-acetylene welding and cutting equipment.

burner greatly exceeds the flame speed the flame will lift off the nozzle. There is a characteristic flame velocity for each fuel gas and mixture of this gas with air or oxygen. This velocity is lowest for the mixtures at the lower and upper limits of inflammability and rises to a maximum in between. These limits of inflammability define the range of gas mixtures where combustion will occur. Acetylene and hydrogen have wide limits of inflammability and high maximum flame speeds of approximately 25 and 40 ft/s respectively. The actual value of flame speed depends on the method of measurement. Propane is inflammable over a narrow range of mixtures and has a lower maximum flame speed of 12·5 ft/s. There is therefore an optimum velocity of gas and flame power. Nozzles for oxy-acetylene welding are frequently graded and numbered according to their heating power in terms of the volume of acetylene in cubic feet burnt per hour at the optimum setting.

The geometry of the orifice in the nozzle affects the above considerations slightly and also influences the shape of the flame, particularly the cone.

The oxy-acetylene flame

This welding flame is produced by supplying nearly equal volumes of oxygen and acetylene to the torch. Primary combustion takes place at the base of the flame in a thin shell-like region which surrounds the cone (fig. 79). The reaction is as follows:

$$C_2H_2 + O_2 \rightleftharpoons 2CO + H_2 + 106\,500 \text{ cal.}$$

Maximum temperature is reached just beyond the apex of this combustion zone, where the gases have a volumetric analysis of approximately CO,

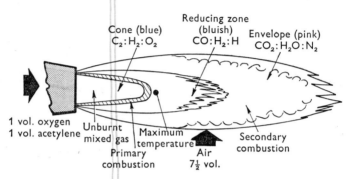

Fig. 79. The structure of the oxy-acetylene flame.

60 per cent; H_2, 20 per cent; and H, 20 per cent. The products of this first reaction form the bluish region of the flame called the reducing zone. Because this region is most closely in contact with the work in welding it largely determines the characteristics of the flame from the welding point of view.

As the burning gases leave the nozzle they begin to entrain air on the outer fringe of the flame. This entrained air mixes with the products of primary combustion and allows secondary reactions to take place. The farther from the nozzle the deeper the entrainment due to turbulence and diffusion. The pinkish-coloured envelope of the flame is therefore oxidizing and has a high nitrogen content as each volume of oxygen in air is accompanied by 3·78 volumes of nitrogen:

$$2CO + O_2 + 3\cdot78N_2 \rightarrow 2CO_2 + 3\cdot78N_2,$$
$$2H_2 + O_2 + 3\cdot78N_2 \rightarrow 2H_2O + 3\cdot78N_2.$$

The chemical characteristics of the flame can be altered to suit the requirements of the welding process by changing the ratio of acetylene and oxygen. For most applications the so-called neutral flame is used. This is essentially the flame as discussed above, but because it is less desirable to have a slightly oxidizing flame than one which is reducing it is usual to arrange for a slight excess of acetylene. Correct adjustment is indicated by a slight white flicker on the end of the cone. Any further increase in the volume of acetylene supplied results in the flame becoming carburizing. Free carbon is produced by the primary reaction and persists throughout the reducing zone. The high temperature renders the carbon incandescent so that the reducing zone becomes luminous and the flame is said to have a white feather. Because carbon remains unburnt from the primary stage the flame temperature drops slightly.

When the proportion of oxygen is increased the carburizing reaction is prevented, the white feather and the reducing zone vanishes and the cone becomes shorter. The usable part of the flame then contains CO_2, H_2O, both oxidizing, and even excess oxygen.

Carburizing flames are used when carbon must be added to the material being welded. When such a flame is used on mild steel the surface layers pick up carbon, resulting in a reduced melting point, so that only the surface fuses. This technique is useful for hard surfacing where deep fusion is to be avoided. Oxidizing flames are used with alloys containing zinc, e.g. brass. The zinc is oxidized on the surface of the pool where the oxide layer inhibits further reaction. With a normal flame zinc is volatilized from the pool continuously and oxidizes in the atmosphere.

Supplies for oxy-acetylene welding

The oxygen for welding is generally supplied compressed to about 2000 lb/in.^2 in steel cylinders. Large users frequently have oxygen delivered in bulk in liquid form and store it for use in refrigerated tanks from which it is evaporated as required. (Similar methods are employed for bulk supply of carbon dioxide and argon used in welding.)

Acetylene is produced by the reaction of water on calcium carbide. The carbide is formed by fusing anthracite or coke with limestone at high temperature in an electric furnace:

$$3C + CaO \rightarrow CaC_2 + CO.$$

After crushing, the calcium carbide is reacted with water to produce acetylene which is then purified to free it from traces of sulphur and phosphorus.
$$CaC_2 + 2H_2O \rightarrow Ca(OH)_2 + C_2H_2.$$

Acetylene can also be made directly from hydrocarbons by several methods, but these are most suitable for large-scale chemical engineering. For the generation of acetylene for the many bottling stations in a distribution system for dissolved acetylene the carbide process is at present more economical.

Acetylene for welding can be supplied in cylinders or generated from carbide and water ready for welding in special plant. Because acetylene is unstable at pressures above 30 lb/in.^2 it cannot be compressed directly into cylinders. Cylinders for acetylene are packed with a porous filler saturated with acetone. The porous mass divides up the space in the cylinder into many small cells making the propagation of an explosion impossible.

Acetone absorbs 25 times its own volume of acetylene for each atmosphere of applied pressure and this permits acetylene to be compressed safely up to 250 lb/in.^{2}. Acetylene treated in this way in cylinders is known as dissolved acetylene. Although dissolved acetylene is convenient to use, some users, though many fewer in recent years, prefer to produce their own supply from calcium carbide and water. This is called generated acetylene, and because the gas is produced at low pressure it requires the use of an injector-type torch in which the mixer is replaced by a Venturi chamber through which the high-pressure oxygen draws the low-pressure acetylene. Injector torches are also used with other low-pressure fuel gases.

Between the supply of gas and the torch it is necessary to have a pressure regulator and a gauge, the two being often combined in one unit. There are many types of regulator, but those for handling high-pressure gases mostly have a basic similarity. They contain a spring-loaded diaphragm the deflection of which closes and opens a small opening connected to the high-pressure side. The pressure of the spring can be varied by an adjusting screw to control the opening of the regulator to give the desired pressure in the exit chamber. Two such devices in series, known as a two-stage regulator, can ensure a constant delivery pressure until the gas in the cylinder is almost exhausted.

Welding techniques

Two different techniques are employed for handling the gas-welding torch. For sheet thicknesses up to $\frac{1}{8}$ in. the forward or leftward technique is used. The torch is held in the right hand and the weld is made from right to left with the torch tilted slightly so that the flame is pointing in the direction of travel (fig. 80). This gives better visibility and directs the flame on to the unwelded metal ahead of the flame. Filler metal is added from the leading edge of the pool. The filler rod size is chosen so that the end can be held within the flame and when brought to the edge of the pool melts rapidly. Generally the same considerations for filler rods apply as for tungsten-arc welding.

With thicker material the backward or rightward technique is used (fig. 81). The torch is still held in the right hand but the weld is made from left to right so that the flame plays on the completed bead. Backward welding concentrates the available heat making the process more suitable for welding thick material. Although gas welding of thick material is slow compared with arc welding there is better control of root penetration so that the process is often used for the first run in pipe joints the remaining filling passes being deposited by metal arc. The flame power required depends on the thickness of the material being welded, its melting point, thermal conductivity and specific heat, also on the welding speed required. For mild steel the flame power in cubic feet of acetylene per hour is approximately 12 for each $\frac{1}{8}$ in. of the thickness.

Process characteristics

The essential process requirements of surface protection and deoxidation are performed mainly by the CO and H_2 in the reducing zone of the flame. For some metals, however, a flux is required in addition. With ferrous materials the reactions

$$FeO + CO \rightarrow Fe + CO_2,$$
$$FeO + 2H \rightarrow Fe + H_2O$$

proceed at welding temperature so that clean welds are produced without a flux. The alumina film on aluminium alloys, however, is not reduced by

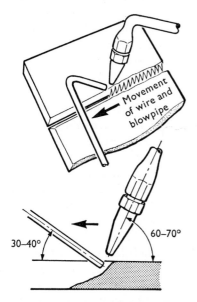 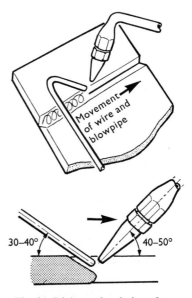

Fig. 80. Leftward technique for Fig. 81. Rightward technique for
gas welding. gas welding.

either CO or H_2 so that a flux must be used with these alloys. Such fluxes are mainly halides of the alkali metals and because they are corrosive they must be removed from the joint afterwards. Designs of joint which might trap flux are avoided. Fluxes are also required with copper and some copper-base alloys and these may be based on boric acid.

Metallurgical control of the deposit is aided by the optimum choice of filler rod. Most filler rods for welding must contain extra deoxidants to control the oxygen content of the molten pool. With steels this is performed largely by silicon, but also by manganese. The correct ratio of these elements is necessary not only to control weld chemistry but also to impart to the pool the most suitable fluidity. During welding the products of deoxidation form a thin film on the metal surface which has a dominant effect on the fluidity and stability of the molten bead. A fluid slag may therefore make positional welding difficult. A rod widely used for mild and carbon steels has the following analysis:

C   0·25 to 0·30 per cent,

Mn   1·2 to 1·50 per cent,

Si   0·3 to 0·50 per cent.

Applications

Gas welding once ranked equal in importance with metal-arc welding. Since the introduction of the inert-gas methods, particularly tungsten-arc welding, with which it has most in common from the view of welding technique, its use has declined for those metals with which a flux is required. It remains an important welding process for steel in sheet thicknesses or where good control, in the root run for example, is required. The equipment is cheap and the process can be applied to a wide variety of metals (pl. 12).

Flames are less efficient and not such compact heat sources as arcs, so that heat-affected zones and weld widths tend to be greater with gas welding. The considerable heat spread with gas welding results in less severe heating and cooling cycles than with arc welding so that gas welding has been widely used for welding hardenable materials such as high carbon and some alloy steels. The spread of heat also results in greater distortion than in arc methods. On sheet metal the process can often match or even exceed the speed obtained with arc processes. There is probably less operator fatigue with the process than with arc methods because lighter viewing glasses can be used and control of torch position is less critical. Because flames are used to heat metals for hot working the oxy-acetylene welding plant is indispensable in the forge and workshop. It is a most useful aid to maintenance welding and is used for brazing and repair work. Automatic oxy-acetylene is successfully used for welding sheet-metal components in which edge joints are used, this making the addition of filler metal unnecessary.

The oxy-hydrogen flame

Hydrogen burns in oxygen according to the reaction

$$2H_2 + O_2 = 2H_2O \text{ (water vapour)} + 2 \times 58\,000 \text{ cal.}$$

The flame so produced is strongly oxidizing and an excess of hydrogen is required to make it reducing and usable for welding. Theoretical flame temperatures cannot be achieved therefore. The hydrogen to oxygen ratio supplied to the torch can vary between 2·5 to 1 and 6 to 1.

The lower temperature of the oxy-hydrogen flame compared with oxy-acetylene has limited the use of this gas mixture, although it gives a clean flame useful for welding and brazing aluminium alloys and the lower melting-point metals. Its stability at high pressures enables the gas to be supplied compressed in cylinders. This stability makes hydrogen necessary for deep underwater cutting where the pressure required to overcome the hydrostatic pressure and operate the blowpipe would make acetylene dangerous.

A recent development has resulted in the introduction of oxy-hydrogen welding for micro-welding and extended its use in the jewellery trade. Oxygen and hydrogen are generated by the electrolysis of water and the mixed gases fed to a miniature torch, the tip of which is a hypodermic needle. The flame power is controlled by varying the electrolyzing current. Such a flame is of course oxidizing and the products of electrolysis can be passed over alcohol to enrich the flame and make it reducing in character. As with the normal oxy-hydrogen flame the excess of reducing agent decreases the flame temperature. The compact unit is operated by mains electricity and only requires water as a fuel which has resulted in the process being given the misleading title of 'water welding'.

ATOMIC HYDROGEN WELDING

When hydrogen is passed through an electric arc the high temperature in the plasma core is sufficient to cause dissociation of the gas. The reaction is endothermic the energy being supplied by the arc

$$H_2 \rightleftharpoons H + H - 100\,700 \text{ cal.}$$

If the stream of gas is now directed against a metal surface the atomic hydrogen recombines giving back the energy taken from the arc. The flame at this point is at approximately $3\,700\,^{\circ}C$ and can be used for welding. In the outer regions of the flame air enters to allow the hydrogen to burn to water vapour in a large woolly flame. Although an arc takes part in the supply of energy, the work is not part of the circuit and the process can be truly described as a flame-welding method in which heat is liberated by a chemical reaction.

The atomic hydrogen process employs a welding torch in which an a.c. arc is struck between two inclined tungsten electrodes. Hydrogen is passed through two nozzles, one surrounding each electrode so that the gas streams converge forming a fan-shaped flame. Because of the high voltage required to ignite and maintain an arc in hydrogen a transformer with the high open circuit voltage of 300 V is required.

The process was widely used for the manual and automatic welding of sheet metal and for such jobs as surfacing dies where the high flame temperature enabled a thin surface layer to be deposited on the thick base metal. Modern inert-gas arc processes have now largely replaced the atomic hydrogen process.

THE THERMIT PROCESS

A number of metal oxides can be reduced by reaction with finely divided aluminium with the liberation of considerable heat so that the products of the reaction are molten. Molten, superheated iron produced in this way can be poured between two parts of a joint to produce a weld. The reaction is obtained with any of the iron oxides but ferric oxide produces the highest temperature, up to 2450 °C being reported:

$$Fe_2O_3 + 2Al \rightarrow Al_2O_3 + 2Fe.$$

A charge of 1000 g of Thermit produces 476 g of slag, 524 g of iron and 181 500 cal. The ferric oxide is prepared from mill scale to which is added other materials to control the reaction of the Thermit and the ultimate analysis of the metal produced. Thermit powder will not ignite below 1300 °C so that the reaction is started with a small quantity of special mixtures and an ignitor. The reaction is completed in about 30 s to only 1 or 2 min regardless of the size of the charge. The quality and soundness of the metal is improved by adding small pieces of scrap steel or alloys to the powder. In this way excessive heating is also avoided and the molten steel produced has a temperature of approximately 2100 °C—that is, a superheat of 600 °C.

The range of metal composition and mechanical properties which can be obtained by additions to the powder and through the scrap metal is considerable. Metal deposited for joining rails, for example, has the following analysis:

C	0·6 per cent	Al	0·4 per cent
Mn	1·4 per cent	Ti	0·04 per cent
Si	0·13 per cent		

All Thermit metal contains traces of aluminium as a residual from the exothermic reaction. Up to 0·7 per cent is useful because as a powerful deoxidant it ensures sound weld metal.

Although the main application of the process is for joining steels, non-ferrous aluminothermic reactions are feasible and copper and copper-based alloys have been used.

Techniques

Thermit welding is suited to the welding of joints with large, compact cross-sections, such as rectangles or rounds. The parts to be joined are cut off square and spaced apart, the surfaces being clean and free from scale. A gap is left between the surfaces to be joined which bears the empirical relationship to the cross-sectional area (A) of $\sqrt[3]{A}/3\cdot6$ in. A mould is then placed round the joint, which has vents, risers and gates as used in foundry practice, but in addition a pre-heating gate (fig. 82). This opening gives access for the pre-heating torch which is used to dry the mould and heat the joint to a suitable temperature for welding as the superheat cannot be

relied upon to provide all the heat for welding. For repetition work pre-
formed moulds may be used, otherwise it is usual to employ a variation of
the lost wax process. The space to be occupied by the Thermit metal is
shaped in wax which is melted out after the mould has been rammed. A
conical crucible positioned over the mould contains the powder, and after

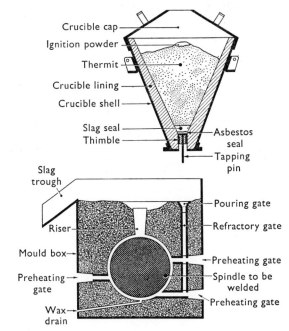

Crucible cap
Ignition powder
Thermit
Crucible lining
Crucible shell
Slag seal
Thimble
Asbestos seal
Tapping pin
Slag trough
Pouring gate
Riser
Refractory gate
Mould box
Preheating gate
Preheating gate
Spindle to be welded
Wax drain
Preheating gate

Fig. 82. Crucible and mould for joining massive sections by the
Thermit welding process.

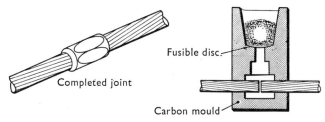

Completed joint
Fusible disc
Carbon mould

Fig. 83. Arrangement for Thermit welding cable with a non-ferrous mixture
using fusible steel disc instead of tapping pin.

ignition the metal settles to the bottom with the alumina slag floating on
top. The metal is then bottom tapped into the mould by knocking up the
tapping pin and the weld is made. With non-ferrous aluminothermic
mixtures permanent carbon moulds have been developed and tapping is
automatic as the charge is arranged to melt through a steel-retaining disc
which replaces the tapping pin (fig. 83).

Applications

The first and now most widely employed use of Thermit welding is for joining rails. It has also been used in heavy construction, shipbuilding, joining reinforcing bars and for repair welding with charges of up to 3 tons. The use of the process in heavy construction is now receiving competition from the more recently developed electroslag process. Non-ferrous aluminothermic mixtures have been used for joining copper conductors.

Bibliography

Ahlest, H. C. W. (1961). Long rails and modern Thermit welding. *J. Perm. Way Instn*, **79**, no. 2, 94–106.
Bainbridge, C. G. (1948). *Gas Welding and Cutting*. London: Iliffe.
Moen, W. B. and Campbell, J. (1955). Evaluation of fuels and oxidants for welding and associated processes. *Welding J.* **34**, no. 9, 870–6.
Seferian, D. (1949). *A Survey of Modern Theory on Welding and Weldability*. (1) *The Welding Flame*, pp. 597–604. Sheet Metal Industries.
The Efficient Use of Fuel (1944). H.M. Stationery Office.
Thermit Welding (1963). Thermit Welding (Great Britain) Ltd., Rainham, Essex, England.

8

RADIATION WELDING

A small group of processes employ energy for welding in the form of radiation. Three processes exist at present based on the optical or arc-image, laser and electron beam systems. The first two processes employ energy as electro-magnetic radiation while an electron beam is a stream of fast-moving particles. It can, however, be classed as particle radiation. Further justification for grouping the processes under the heading 'radiation welding' is found in an examination of their welding characteristics.

Radiation welding methods are unique in that the energy for welding may be focused on the object to be welded, heat being generated only where the focused beam strikes the workpiece. Unlike arc or flame sources, therefore, the work is not brought in contact with any heated media, gas or metal vapour, and the processes may be carried out in vacuum or low-pressure systems where the ultimate in cleanliness can be achieved. Finally, in contrast to arc welding, the melted pool is subjected to only negligible pressure.

ARC-IMAGE WELDING

Fusion is accomplished by focusing the image of a high-temperature source on the workpiece. Mirrors are used as they are more practical than large quartz lenses which would otherwise be required and a large aperture is necessary so that the maximum energy may be collected.

In the practical utilization of the method fundamental difficulties are encountered. The small high intensity spot which is ideal for welding necessitates the use of a point source emitting radiation of high infra-red content. As a point source does not exist, the most compact source available must be selected, the main requirement being a high intensity rather than a high total output. This is achieved more readily as the temperature of the source is raised because of the dependence of radiation on the fourth power of the temperature. A high temperature must also be used because it is fundamentally impossible for incoherent light to be focused to an image hotter than the source. High-pressure plasma-arc sources have been developed with outputs well above 10 kW which have been used for brazing and welding. Optical systems with top surface mirrors of high accuracy are required for focusing, but even so losses must occur by dispersion because the source is finite and emits in all directions. The resultant image has blurred edges and the energy distribution at the work is non-uniform with its maximum level well below that of the source. Furthermore, a high proportion of the incident radiation may be reflected from the work surface. Although the method has no future for terrestrial welding it may eventually be employed in space, using the sun as the heat source.

LASER WELDING

The laser is a device which when irradiated by light from an intense source is capable of amplifying radiation in certain wavebands and emitting this as a coherent parallel beam in which all waves are in phase. This beam can travel long distances without attenuation and may be focused through lenses to produce spots in which the energy density is equalled only by the electron beam. Because the radiation is coherent, energy concentration is feasible in a way not possible with the arc-image system. The name laser is taken from the initial letters of the type of radiation which makes this possible—light amplification by stimulated emission of radiation.

Above the temperature of absolute zero atoms can exist with different well-defined levels of energy. Atoms may be moved to higher levels by absorbing radiation and then return either directly or through intermediate steps to their initial state releasing energy at each step. To satisfy the condition of thermal equilibrium the number of atoms present in each state of energy must be less than that of the next lower-energy level or the ground state. Radiation passing through material in this condition is attenuated. If the material can be brought into a condition in which the population of atoms at the higher-energy level is actually greater than the lower level (a population inversion), however, the radiation is amplified. This amplification is effective for a narrow line width given by Bohr's frequency relation $hv = E_2 - E_1$, when E_1 and E_2 are the energies of the states between which the transition is taking place, h is Planck's constant and v is frequency.

Population inversion can be made to occur in solids and gases. One of the most common laser materials and also one capable of delivering high power is the chromium ion in a ruby crystal. A typical concentration would be 0·05 % Cr. Another solid-state laser material is glass containing 2–6 per cent of neodymium. A practical form of laser might consist of a rod of ruby 1 cm diameter and 10 cm long with accurately ground and polished ends, one of which is half silvered the other fully silvered. The crystal would be irradiated by the light from a powerful xenon flash tube as shown in fig. 84. Within the crystal chromium ions emit stimulated radiation and that travelling axially is reflected to and fro between the ends of the crystal. At each reflection a certain loss occurs, but if the amplification of the radiation during transit is sufficient to leave a net gain of energy, the intensity of radiation builds up and the laser beam is radiated from the half-silvered end. The process only takes place therefore if a certain threshold of input energy is exceeded in the presence of feed-back within the resonant cavity formed by the crystal. Radiation emitted in this way is essentially mono-chromatic, its frequency depending on the active (e.g. the chromium or neodymium) ions and the energy levels stimulated.

The high-energy levels necessary to cause laser action can usually be achieved only intermittently by discharging a bank of capacitors through a xenon flash tube—a process known as pumping. These capacitors are charged to between 1 and 4 kV. The light intensity of the flash tube rises to a maximum within 1 ms and decays exponentially over a period up to 10 ms. Fluorescence begins almost immediately within the crystal, but after an interval of $\sim$ 5 ms stimulated emission of coherent light occurs in

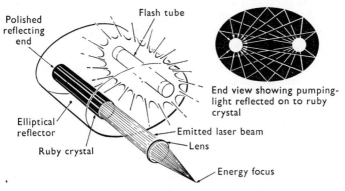

Fig. 84. Principle of the laser power source.

a series of spiky pulses. An overall efficiency of less than 1 per cent is obtained with an actual output of frequently between 2 and 50 joules. Intermittent operation is a result not only of the inability to supply the high level of input power continuously but also because the process itself generates heat within the crystal which must be allowed to cool as an increase in tem-perature raises the threshold condition. A repetition rate of several times per minute is the most that can be achieved without elaborate cooling procedures.

In spite of the modest total energy the short period of each pulse and the ability to focus the beam to a small diameter allows energy concentrations of over 10^8 W/cm^2: a level more than adequate to vaporize metals. When operated at these high-energy densities the process can be employed for drilling holes and cutting. For welding, however, the energy should not arrive faster, or much faster, than it can be conducted away as heat. This requires a less spiky, more sustained pulse which is achieved by increasing the inductance of the flash tube circuit or employing a delay line.

The laser has been used for micro-spot welding and this appears to be its most promising application, particularly where minimum spread of heat is required. Until a continuously pumped laser of adequate power has been developed, however, it will remain a single-spot process and seam welds must be made by overlapping individual spots. Unlike the electron beam process a vacuum chamber is not required for the generation and delivery of the beam. When micro-welding, the high rate of delivery of energy and the brief thermal cycle could make precautions to prevent oxidation during welding unnecessary.

ELECTRON BEAM WELDING

A welding process involving melting in which the energy is supplied by the impact of a focused beam of electrons. Important features of the equipment for electron beam welding are: the electron gun, in which the stream of electrons is produced and accelerated; the focusing and beam-control system; and the working chamber, which must operate at low pressure. This latter is required to permit the generation and passage of the electrons and to prevent damage by contamination to the heated cathode or filament. Operation in a vacuum is an incidental advantage when welding many metals sensitive to contamination by the atmosphere.

Types of electron gun

There is a variety of types of electron gun, a reflection of the fact that electron gun design is a compromise with differing solutions according to requirements. Electrons are produced by a heated filament or cathode and are given direction and acceleration by a high potential between the cathode and an anode placed some distance away.

In a limited number of electron beam-welding plants the anode is the work itself. This system is known as 'work-accelerated' and, although simple, has a number of limitations as far as welding is concerned. Fine focus of the beam is difficult, the work must be electrically conducting, the gun must be used close to the work, and the geometry of the work and its surface affect both the focus and control of the beam. Work-accelerated guns have been used where large numbers of a particular component of simple shape have to be welded; for example, reactor fuel elements. Because of their many limitations this type of gun is of little significance in welding, the more adaptable 'self-accelerated' gun being invariably employed.

If, instead of making the work the anode, a perforated plate is substituted, the electron beam will pass through this plate and can be made to impinge on the work beyond. The gun is then said to be 'self-accelerated'. As the anode may be arranged near the first cross-over the electron beam is divergent as it approaches the work and a lens system is required between anode and work to bring the beam to a focus. This is normally achieved with an electro-magnetic lens comprising a coil encased in iron with a circumferential gap in the bore. The iron concentrates the field and allows shorter focal lengths than would be possible with an unshielded coil. On entering the magnetic field of the lens the individual electron paths are both rotated and refracted and they emerge as a convergent beam. The focal length of the lens may be adjusted by varying the coil current. Pairs of 'deflection' coils placed at right angles after the lens system allow the position of the focal spot to be manipulated in the work plane normal to the beam. In this way fine adjustments may be made in the position of the spot or the spot may be programmed electronically to weld a complex shape.

Two popular types of electron beam-welding gun are illustrated diagramatically in figs. 85 to 87. The Pierce-type gun shown in fig. 85 behaves

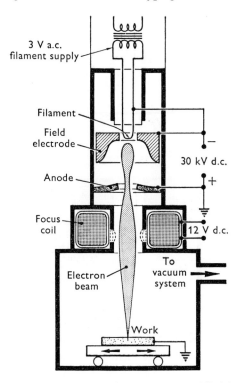

Fig. 85. Pierce-type electron beam gun (diode).

essentially as a diode valve in that the current (I) passed by the device is limited by the space charge. This is shown by the static characteristic curve relating I and V, the accelerating voltage. The curve can be described approximately by the relationship $I = AV^{\frac{3}{2}}$, and the ratio $I/V^{\frac{3}{2}}$, known as the perveance of the device, depends on the geometry of the gun. For each acceleration voltage there is only one beam current unless the geometry of the gun is altered. Pierce-type electron guns are made therefore so that their configuration—particularly the anode/cathode distance—can be altered to provide changes in beam current. The alternative of varying the accelerating voltage is not acceptable because of practical difficulties. The shaped electrode in the plane of and at the same potential as the cathode of the Pierce-type gun has the function of providing a field which converges the electron stream.

In other types of electron gun resembling the triode thermionic valve the beam current is controlled and varied by a negative bias on an electrode placed round the cathode which is called the control electrode or bias cup.

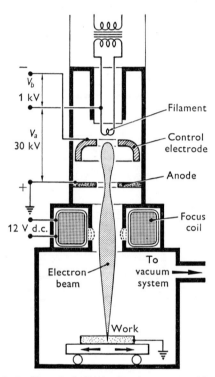

Fig. 86. Outline of triode electron beam gun with control electrode or bias cup.

This electrode behaves as a grid in that it is able to influence the space charge and thereby control the flow of electrons from the cathode, but it also exercises an influence on the equi-potential planes along the axis of the gun. A deep cup-shaped electrode is particularly effective in giving the stream of electrons an inward component of velocity so that the beam comes to a focus. The negative bias on the control electrode is within the range 0–1 000 V and, since a supply of this voltage can be made infinitely variable, adjustment of the bias voltage is a convenient method of controlling the beam current. Changes in gun geometry are not required. Triode valves have a series of characteristic curves, each describing the volt/amp relationship for a particular grid voltage. The more negative the grid is made with respect to the filament the more it assists the space charge in repelling electrons from the filament and the lower is the current passed through the device. Changes in the potential of the control electrode influence beam focus.

Spot size

In practical forms of electron beam-welding gun with electro-magnetic focusing the accelerating voltage may range from 15–150 kV and the cur-

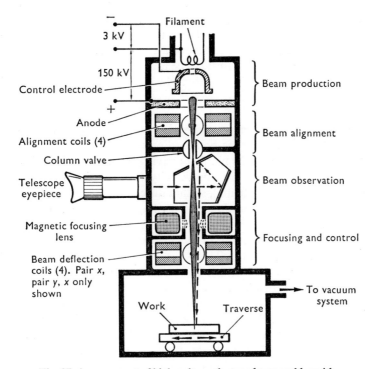

Fig. 87. Arrangement of high-voltage electron beam welder with coaxial optical viewing.

rent carried by the beam up to 0·5 A. The minimum focal spot size is between about 0·005 and 0·025 in. diameter depending on the power level and the accelerating voltage as well as on the design of the gun.

Spot size is an important factor in welding as it controls the width of the weld directly and through its effect on power density, also the ratio of width to penetration. There are several reasons why a small spot size is difficult to obtain. Electrons leave the cathode surface with a range of initial velocities and in all directions. Ideally therefore the cathode should be kept as small as possible consistent with its ability to pass the re-quired heating current. Unfortunately the highly efficient doped cathodes employed in valves cannot generally be used for welding because of the imperfect vacuum conditions inherent to welding and so tungsten filaments are usually employed as they are less sensitive to contamination. Precise control of cathode geometry is important as distortion of its shape during use can change the paths of the electrons which leave its surface. In a gun with a cathode heated directly by the passage of an electric current the magnetic field produced by the heater current may influence the electron paths. Finally, the space charge effect, the mutual repulsion of particles of like charge, increases the minimum spot size which can be achieved as the charge density (or beam power) is raised because it becomes more difficult to hold the beam together. High voltage and low beam current tend, therefore, to favour a small spot size. It will be clear, however, that the beam most likely to be useful for welding—long, narrow, dense and finely focused—is unlikely to be obtained without compromise.

Beam power

The heat in electron beam welding is obtained by the release of the kinetic energy of the electrons bombarding the work and depends therefore on their velocity and the number arriving in unit time. Kinetic energy is given by $\frac{1}{2}mv^2$, but v, the electron velocity, is proportional to the square root of the accelerating potential, so that the energy per electron is directly proportional to the accelerating voltage. Since the number of electrons arriving per unit time is directly proportional to the beam current the beam power can be described in terms of the product of accelerating voltage and beam current, that is, in watts. This value divided by the spot size gives the energy density at the work which can amount to 5×10^8 W/cm^2, a significantly higher figure than for any arc welding process.

Methods of varying beam power have been discussed under the heading of electron gun types. Where electron beam-welding equipment uses bias control of beam current the focus and spot size is changed as well, because of the space charge effects, and adjustment of the focus coil current will be required with every alteration of beam power. When welding a workpiece of varying thickness which would require different beam powers the procedure can become complicated. Electronic coupling between the two controls has been attempted to make these adjustments automatic. Another solution is the constant-focus Pierce-type gun with beam-power control by modulating the accelerating voltage, a system under development.

Operating voltage

The same beam power could be obtained with an infinite range of beam current and accelerating voltage, but in practice guns tend to be of two types: high voltage, 70–150 kV and low voltage, 15–30 kV. The reasons for this have to do with both the generation of the beam and the type of work it is to do. High-voltage systems allow finer spot sizes, deeper penetration, longer focal length and greater working distances. The guns tend to be long and with the high-voltage insulation required must remain stationary which means that the work must be moved under the gun. Because the beam is long and narrow, accurate positioning is required on the work and this necessitates the use of a coaxial optical viewing system (fig. 87). The above limitations need not be disadvantages and indeed the high-voltage system is unrivalled for precision work and where welds must be made with restricted access.

With guns of the same beam power but lower accelerating voltages the working distance tends to be shorter and the beams more convergent to limit space-charge effects. The position of the work with respect to the focus is therefore more critical than with high-voltage guns. If such a gun were stationary the working area it would command would be reduced so that low-voltage guns are frequently designed to be used inside the chamber and are themselves moved around the stationary work. This practice permits a considerable reduction in vacuum chamber size. The same could not be done with high-voltage guns because their greater bulk and more complicated insulation limits their portability.

When electron beams are intercepted X-rays are generated. The emission depends on the square of the voltage but is directly proportional to current. Neither high- nor low-voltage systems present insuperable problems in protection, but with less than 30 kV the problem is greatly simplified.

Pulse techniques

When metals are heated in vacuum, absorbed surface films and gases may be driven off which can degrade the vacuum. Contamination of the electron source may then occur or gas discharges take place within the gun. Volatile metals can cause similar troubles. Electron guns vary in their susceptibility to the effects of out-gassing according to their design. One method of minimizing the effects is to operate the electron gun intermittently by pulsing, with the object of allowing gas to disperse harmlessly in the brief periods while the gun is switched off.

Pulse techniques are also used to influence penetration and control heat flow and heat-affected zones. With deep penetration techniques vapour in the crater can de-focus the beam. Intermittent operation of the beam prevents this by allowing dispersion of the vapour between pulses. By operating intermittently at high-beam power, the temperature gradient can be increased over that obtained with the equivalent continuous power.

Deep penetration

A most attractive feature of electron beam welding is the ability to produce welds with extraordinarily deep penetration ratios, values of up to 20:1 being reported. Deep narrow penetration can also be obtained with an electric arc system. It will be recalled that with both inert-gas tungsten-arc and inert-gas metal-arc welding a 'finger' develops in the penetration when the welding current rises above 350 A. This is because a higher degree of ionization is achieved and the arc core becomes more concentrated, with an associated increase in temperature and pressure of the plasma stream. The plasma stream creates a deep crater in the liquid pool thereby allowing the high-temperature gas streams to impinge directly on the solid metal beneath the pool. In the extreme condition the molten metal is either expelled completely from the pool or a cutting arc is produced. As neither condition is suitable for welding no attempts have been made to utilize a deeply fingered penetration in arc welding (fig. 88 *a*).

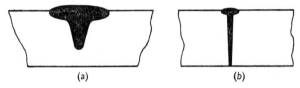

(a)   (b)

Fig. 88. (*a*) Finger penetration in a gas metal-arc weld. (*b*) Deep penetration in an electron beam weld.

Finger-type penetration, often called 'nailhead' penetration in the electron beam context, is also observed when the power density of the electron beam is increased above a limiting value. With electron beam welding, however, the pressure due to bombardment by the electrons is negligible because of their low mass, and gas streams are absent so that it is possible to achieve high temperatures without the excessive pressures associated with welding arc. Deep penetration welding is therefore feasible (fig. 88 *b*).

Electron beams themselves penetrate metals to only a limited extent depending largely on voltage as given by the relationship

$$P = \frac{\text{At. wt. } V^2}{\text{At. no. } \rho K},$$

where V is the accelerating voltage, ρ density and K a constant. This penetration, called the electron range, is insufficient to account for the finger penetration, which is a mechanism that appears to depend on volatilization. Where the power density of an electron beam is such that energy is arriving faster than it is conducted away as heat, the temperature rises until volatilization occurs. A crater then develops because of the removal of metal and the reaction pressure exerted by the vaporizing metal leaving the crater. Metal vapour in the crater does not scatter the beam, in fact there is some evidence that it can assist in focusing the beam. The crater continues to grow in depth and, as with an accurately focused beam, there is only limited spread of heat to the walls when the beam is stationary, the liquid metal lining the crater is held in place by surface tension and vapour pressure. When the beam is traversed the crater moves forward and the liquid metal at the leading edge flows behind, where it solidifies. The relative importance of vapour pressure and surface tension in the process depends on the size of the hole. Reaction pressure is more important with larger holes. As the deep penetration-welding technique depends largely on vapour pressure and surface tension, it is found that metals vary in their response to this type of welding. Stainless steel is easily welded with deep penetration, zirconium is less suitable. Where a beam penetrates through a workpiece completely the crater is open at the root so that vapour can pass both ways. This condition is frequently associated with fine spatter on the reverse side of the joint.

Although the electron range in solid metals is normally negligible it can be significant when welding thin films on glass or ceramic substrates, because considerable heat can be liberated in the substrate. The range of the light waves used in laser welding is orders of magnitude less than electron beams so that, although the processes can deliver comparable energy densities, they may have different properties as regards penetration.

The direction of solidification in welds with a high penetration ratio frequently appears to be favourable to the development of a line of central weakness. That no such weakness has been reported may be because of the clean condition of the weld pool and that there is no gas, slag or foreign material to trap between the opposing dendrites. The thermal cycle is also rapid so that the crystal size is smaller than in the deep-penetration arc-welding processes.

Applications

The high intensity of the electron beam as a heat source results not only in deep narrow penetration but also in the associated effect of narrow heat-affected zones. Both effects together are of considerable importance as they result in reduced distortion and improved mechanical properties in joints when compared with other welding processes. For these reasons the process is not limited to the welding of those metals such as zirconium and molybdenum which, because of their high reactivity, are best welded in a vacuum. The necessity to use a vacuum is at once both an advantage and a disadvantage. It enables fusion welds of unequalled quality to be made but the size and shape of the workpiece is limited by the vacuum chamber. The combination of vacuum and high temperature ensures the breakdown or volatization of surface films giving clean fluid weld pools.

Many chambers contain elaborate mechanisms for manipulating the work and inevitably electron-beam welding equipment is expensive and tends to be specialized. Attempts have been made to devise methods of passing the electron beam into a gaseous shield so as to dispense with the chamber. While limited success has been achieved the beam is greatly degraded by scattering as a result of collisions with the molecules of gas and has limited range and, of course, the quality advantage of welding in vacuum is sacrificed.

A variety of joint types can be welded by the electron beam. Because of the high speed attainable, however, and the difficulty of feeding filler wire remotely, it is general to use close butt edge preparations which do not require filler. Alternatively, the preparation may be machined with a fin to provide extra metal. Joints must be machined and fitted accurately. This does not usually present difficulties as, due to the lack of distortion and cleanliness of the process, welding is often done on finished or semi-finished machined parts. As shown in fig. 89 three types of joint are unique

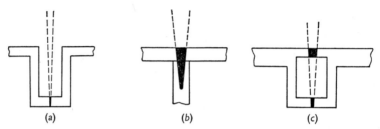

(a)   (b)   (c)

Fig. 89. Types of joint specifically applicable to electron beam welding. (*a*) Welding in a recess. (*b*) Welding from outside with the 'spike' technique. (*c*) Simultaneous multiple weld joint.

to electron beam welding: (*a*) with beams of narrow convergence it is possible to weld in inaccessible positions, e.g. in the bottom of a deep narrow cavity; (*b*) the deep penetration technique permits the making of a fillet joint from the outside using what is called the 'spike' technique, and (*c*) deep penetration with high-density fine beams also permits two or more welds to be made at once, one above the other.

Bibliography

Bakish, R. (ed.). *Introduction to Electron Beam Technology*. John Wiley and Sons.

Kaehler, W. A., Bank, S. (Jnr.) and Trabold, A. F. (1965). Arc-image welding. *Weld. J.* **44**, no. 11, 497S–503S.

Platte, W. N. and Smith, J. F. (1963). Laser techniques for metal joining. *Weld. J.* **42**, no. 11, 481S–9S.

Proc. of 4th and 5th Annual Meetings, 1962 and 1963 of Electron Beam Symposium. Alloyd Electronics Corp.

Sayer, L. N. and Burns, T. E. (1964). Practical aspects of electron beam welding. *Br. Weld. J.* **11**, no. 4, 163–71.

Schmidt, A. O., Ham, I. and Hoshi, T. (1965). An evaluation of laser performance in microwelding. *Weld. J.* **44**, no. 11, 481S–8S.

9

SOLID-PHASE WELDING

Processes in which welding is carried out in the solid phase are less numerous and more highly specialized than those in which union is effected through a liquid phase. There are, nevertheless, a number of such processes which together can be used to cover a wide range of applications.

Because the mechanism of pressure welding depends fundamentally on the properties of metal surfaces it has been the subject of considerable research and undoubtedly has many features in common with friction and wear. Although some aspects of the process still require examination much is now known about the major parameters. In the simplest terms, all solid-phase welding methods depend upon the rupture or dispersion of surface films and the union under pressure of the virgin metal surfaces produced. Rupture of the surface films of contaminants is achieved by the deformation or extension of the interface and this is usually brought about by the bulk deformation of the parts to be welded, as for example in forge welding. It is possible, however, to achieve the same result of surface extension without bulk deformation by employing localized or interfacial deformation. Ultrasonic and impact explosive welding depend on interfacial deformation.

With cold pressure welding and all solid-phase processes there is a minimum interfacial extension and hence deformation below which welding does not occur. The reason for the existence of this threshold deformation is the main point of difference between the several theories of pressure welding which have ascribed it to the rupture or dispersal of surface films: the plastic deformation of asperities; elastic recovery of deformed regions; and the energy required to reorient the structure at the interface. Although these factors are usually discussed separately there does not appear to be any reason why all these theories should be mutually exclusive. The threshold deformation is lowered as the welding temperature is raised. Welding temperature in fact is a most important parameter in solid-phase welding as heat drives off absorbed and chemi-sorbed contaminating surface films, renders the parent metal more plastic, promotes metallic and oxide diffusion; and if high enough recrystallization as well. The temperature of the workpieces can be raised by electric, chemical or mechanical processes.

In considering the effect of temperature on the weldability of different metals the melting point should be taken into consideration. It has been pointed out that compared with lead at room temperature red hot tungsten is 'cold'. A basis for comparison exists in the homologous temperature, i.e. the welding temperature divided by the melting point, both in degrees Kelvin. Milner and others find that for the ductile cubic metals a welding temperature exceeding half the melting point (°K) results in greatly enhanced weldability with minimum deformation.

In practice solid-phase welding is generally done without atmosphere control as, by the nature of the process, the workpieces are pressed together excluding air. This is not completely efficient so that when welding at elevated temperature a reducing atmosphere or vacuum might be employed to assist oxide removal or an inert-gas shield to protect reactive metals.

COLD-PRESSURE WELDING

Where welding is accomplished at ambient temperature solely by the application of pressure across the interface the process is termed cold welding. Cold welding is applied particularly to the ductile metals aluminium and copper, although ductility in itself is not the sole criterion of weldability. Weldability decreases with increase in hardness and melting point, silver for example, although a ductile metal being less readily welded than aluminium. The joints produced are of two types: lap and butt. With the former, indenting dies may be forced into the metal causing deformation and flow to provide the extension of the interface. Roll bonding is a specialized form of lap welding in which behaviour is slighly different in quantitative terms from welding with indenters. The butt method is used for joining wires and bar stock, the parts being gripped in dies and forced together to cause lateral flow.

Weld formation

Surface preparation is probably the most important single process variable. In lap welding the preferred method is scratch brushing after degreasing. Surfaces baked at high temperature are also suitable for welding. However, both scratch brushed and baked surfaces must be welded as soon as possible and not be subjected to any further treatment. Anodized aluminium can frequently be welded without preparation.

With butt welds between bar stock or tubes scratch brushing is not normally practical and it is usual to file or shear the edges square immediately before welding.

Oxide films are not necessarily a barrier to welding, they may even be of assistance with aluminium and copper. The two oxide surfaces on each side of the interface lock mechanically and behave as one, breaking up when deformed to expose fresh metal surfaces which bond readily. A scratch brushed surface behaves in a similar way. The scratch brush treatment removes the surface contaminants, including water vapour, but also produces a hard heavily worked and oxide-impregnated layer which promotes welding by rupturing in a brittle manner. Water vapour or contaminating films can spread from the oxide to prevent metallic bonds forming and because unlike the oxide these films may be extended without rupture their effect can seriously weaken joints. Films such as these may be important in establishing the threshold deformation.

A welded interface comprises, therefore, discrete areas where adhesion is prevented by oxide films, separated by a network of metallic bonds. Assuming that the two oxide films behave as one and that they have zero ductility it has been shown that the maximum area of metallic bonding can be estimated from the deformation during welding. The actual strength is higher than would be judged from this area, however, because of the effect of three-dimensional restraint across the narrow necks where bonding has occurred. On this basis a formula has been arrived at by Vaidyanath relating strength to deformation for cubic metals

$$\frac{\text{Ult. shear strength weld}}{\text{Ult. shear strength solid metal}} = R(2 - R),$$

where R is the % reduction. For hexagonal metals the relationship is different because there is some evidence that the oxide films do not behave as one. There is also an implicit assumption that the bond between the two ruptured surface films has no strength. This may not always be true as with scratch brushed films there is the possibility of bonding between the asperities on the two surfaces especially when elevated temperatures are employed.

Where dissimilar metals are joined the deformation may be different on opposite sides of the interface. The relative movement of the interface improves the welding and reduces the threshold deformation. Heat treatment of the joint after welding produces a number of effects dependent upon the characteristics of the two metals. Metals which are metallurgically immiscible may be softened, but the joints otherwise unaffected. With combinations in which diffusion can take place porosity may occur at the interface if the rates of diffusion of the two metals in each direction across the interface are different. Micro-voids so formed can grow by coalescence into large pores. Diffusional porosity may possibly be reduced by the use of high-purity metals in which the sites of nucleation are fewer. Voids resulting from gas trapped at the interface may be reduced by employing higher pressures and softer materials. If the combination of metals is such that intermetallic compounds are possible a brittle layer can be formed which will grow in thickness in a time–temperature dependent manner. Thick brittle films can cause destruction of the joint.

Techniques for lap welding

Much of the success of cold welding depends on the design of the dies or indenting tools used for welding. For lap welds in sheet indenting dies with flat surfaces have been found most satisfactory. As the indenter is forced into the sheets they tend to separate slightly. The actual weld is thus smaller than the indenter face and a dead area, unaffected by metal flow, develops next to the face of the indenter (fig. 90). With this simple type of die the weld strength can be increased within limits by increasing the area of the indenter as this increases the area of bond.

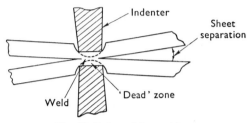

Fig. 90. Indented lap weld.

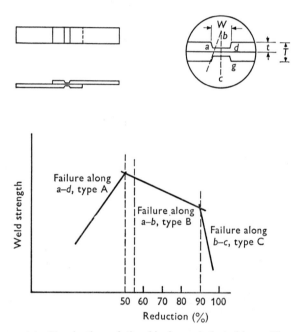

Fig. 91. Strength/weld reduction relationship for an indented lap weld.

With any particular size of spot-weld indenting die the joint strength increases with depth of indentation until a maximum is reached and then further indentation causes a reduction in strength. The shear strength/indentation curve is characteristic for each material and the minimum deformation is an indication of weldability. This tendency for a maximum in the curve is a result of changing modes of failure in shear testing (fig. 91). Strength increases with indentation as the bond area increases, then begins to fall as the greater reduction permits failure through the edge of the weld and finally falls rapidly at high reductions, the strength being dominated by the residual thickness. The shape of the strength/indentation curve is affected by the width of the indenter. An increase in width gives higher strength at lower indentations but the curve is compressed, becoming more peaky, indicating a more critical welding condition. Indenters are usually between one and three times as wide as the single-sheet thickness and six times as long.

A considerable increase in spot-weld strength can be achieved by using a double-acting die. If the indenter is surrounded by a collar which grips the sheets before the weld is made separation of the sheets is impossible. There is then a lateral movement of metal at the interface with high indenter penetration, e.g. over 70 per cent, which results in an increase in weld

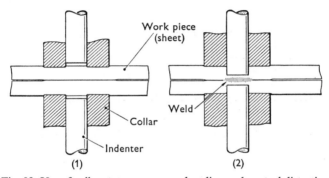

Fig. 92. Use of collars to cause corona bonding and control distortion.

area by corona bonding as shown in fig. 92. A similar effect is noticed in the resistance spot welding of materials such as aluminium-coated aluminium alloys.

Many of the foregoing remarks on pressure-spot welds can apply equally well to seam welds made with rolls having indenting ribs. It is also possible to weld and shear in the same operation. This technique, which is used for longitudinal welds in tube made from formed strip and in canning, relies largely on corona bonding. Figure 93a and b illustrates one design of die for canning.

Techniques for butt welding

Butt welds are made by clamping the two parts to be welded in split dies. Special surface preparation is not required if the ends to be joined are freshly cropped. Each part projects beyond the face of the die so that as the dies are forced together the metal between the dies is extruded laterally to provide the deformation necessary at the interface. The dies for butt welding may have flat or conical faces and the latter have two advantages—

the welding force is greatly reduced and the extruded metal is sheared off as the dies close. A suitable angle for the faces is a cone of 120° (fig. 94 *a*, *b*).

The deformation depends on both the design of the die face and the projection of the parts beyond the die face. With conical-faced dies which close, the deformation is directly related to projection. Round bars re-

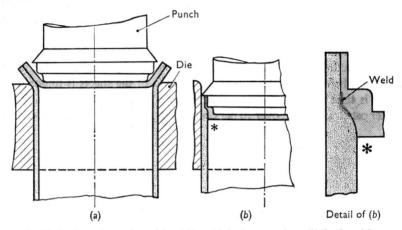

Fig. 93. Sealing of cans by cold welding: (*a*) before pressing; (*b*) final position.

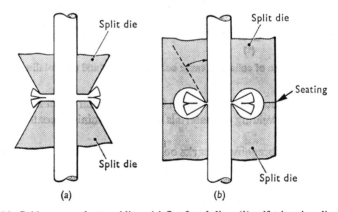

Fig. 94. Cold-pressure butt welding: (*a*) flat-faced dies; (*b*) self-trimming dies.

quire a projection on each side of the interface of not less than 40 per cent of the diameter. With rectangular sections the projection is related to the smaller dimension and is approximately 40 per cent for square bars but rises as the difference between the two dimensions increases.

Where dissimilar metals are cold-welded the softer deforms first and as deformation rises the harder metal begins to flow. The projection of the softer metal must be slightly greater than when similar metals are joined and considerably greater for the harder metal. A common dissimilar combination is aluminium to copper where the projection of the copper must be 30–40 per cent greater than for the aluminium.

Applications

The decision to use cold welding must be taken early in the design stage of a component because allowance must be made for the deformation, and the design must permit the specialized form of the joint. Large numbers of similar joints must be required because dies must be made. Lap welds are used for can joints, longitudinal tube joints and electrical connections; butt joints for wires and tubes. Hand tools are used for small sizes, power-operated presses for butt joints up to about 1 in.^{2} in aluminium. The most commonly welded metals are aluminium and copper.

THERMO-COMPRESSION BONDING

A form of pressure welding done at slightly elevated temperatures has been developed for the miniature joints in the electronics industry, particularly those between fine wires often of 0·001 in. diameter and less and metal films on glass or ceramic. Deformation is carried out at temperatures from 200 °C upward. There are several versions of the process, three being chisel or wedge bond, nail-head bond, and electric resistance or parallel-gap bond. In the wedge bond (fig. 95 a) the wire is deformed against the film by the pressure of a wedge-shaped indenter. The film surface may be heated to 200–350 °C or the wedge itself may be heated. Nail-head bonding is done by threading the wire through an indenter and parting it with a micro-hydrogen flame. The globule left by the parting operation is deformed against the film by the pressure of the indenter (fig. 95 b). Heat is applied to the film and substrate. In the electric resistance or parallel-gap bond the fine wire or strip is pressed to the film by twin electrodes often of high resistance material such as tungsten. The passage of current down the electrodes and through the strip raises the temperature of the strip locally because of contact resistance, and the high resistance of the electrodes prevents loss of heat through conduction. As times of only a few cycles are used the heat input is limited and it is possible to raise the temperature locally to almost the melting point of the wire which aids welding without damaging the electronic device in which the welding is done (fig. 95 c). All three methods can be used with local inert atmospheres. The role of temperature in the process is twofold: to disperse surface films and possibly allow recrystallization. As the recrystallization temperature is exceeded both the threshold deformation and pressure required drop markedly. The noble metals and aluminium and copper are frequently welded by this process and as dissimilar combinations are often involved in thermo-compression bonds diffusion and alloy layer growth phenomena can occur. Parallel-gap joining can also make a brazed joint in coated metals or even a series-type resistance spot weld, although the latter can result in splashed or faulty joints.

GAS-PRESSURE WELDING

Pressure welding at elevated temperatures in which the joint area is heated by flames is generally known as oxy-acetylene pressure welding; however, other fuel gases may be used. The process has been used for a variety of shaped joints but, because torches must be designed to give uniform heating, there is a preference for simple shapes such as bars and tubes. Split ring-type burners are frequently used, and to assist in uniform

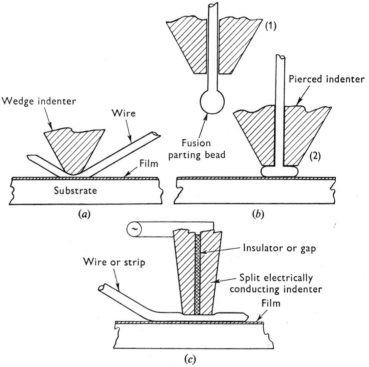

Fig. 95. Thermo-compression bonding techniques: (*a*) wedge bond; (*b*) nail-head bond; (*c*) resistance or parallel-gap bond.

heating these burners can be oscillated slightly. Parts to be joined are machined with square faces and butted together under loads to give for mild steel a pressure of about 1 000 lb/in.². As the temperature builds up deformation takes place and, when the joint area becomes sufficiently plastic, an upset force of about four times the initial force is applied. The heating time for a given material depends on the metal thickness, the burner power and the burner-to-work distance. For mild steel the heating time would be about $4\frac{1}{2}$ min/in. of thickness. Because in gas-pressure welding the area heated is broad the deformation produces a smoothly contoured thickening at the weld area. This thickened, or upset area, is generally about 50 per cent longer than the material is thick. Too small an upset indicates inadequate heating and the reduced deformation may result in a poor weld (cf. fig. 76, p. 138). Excessive deformation can result from overheating. It should be noted that upset in itself is not a criterion of weld quality and that what matters is the deformation at the interface. A narrow heated band will give greater interfacial deformation for a given upset than a wide heated band. There is an advantage then in increasing the heat input, but this is limited by the thermal diffusivity of the material as only the outer surface receives heat and must not be melted (fig. 96). To obtain a well-shaped joint and avoid overheating the outside, a chamfer with an included angle of approximately 120° may be used.

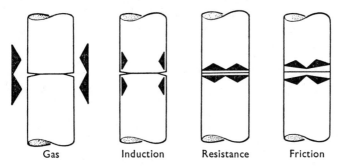

Gas   Induction   Resistance   Friction

Fig. 96. Summary of thermal conditions in pressure-butt welding processes. Arrows indicate source and direction of heat flow.

The function of the initial force is to seal the interface from the atmosphere. Accurate mating of the surfaces even to the extent of grinding is therefore important, particularly with high carbon steels where decarburization may occur or with materials which form refractory oxide films. For the same reason higher initial forces than with mild steel are required for high carbon, alloy steels and non-ferrous metals. The force may be maintained throughout the whole welding cycle, a technique referred to as constant pressure welding, or the pressure may be relaxed slightly once heating begins (to reduce deformation) and reapplied at the end of the heating period. This is a triple-pressure cycle. It may also be necessary with the materials mentioned above to assist bonding by increasing the deformation at the interface by increasing the upset.

Temperatures should be high enough to give rapid recrystallization and adequate plasticity. Those materials which absorb their own oxides at elevated temperature, such as mild steel, are more readily welded. Uniformity and consistency of temperature tend to be difficult to obtain in spite of automatic controls, and it is possibly because of this that the gas-pressure welding process is not widely used.

INDUCTION-PRESSURE WELDING

In this process low-frequency induction heating is substituted for oxyacetylene flame heating. An important difference between flame and induction heating is that in the latter heat is generated by the circulation of induced currents within the metal (fig. 96). Induction heating is therefore rapid but can be controlled well, and special shielding atmospheres can be used if required.

Uniformity of heating throughout the joint is required so that the power source frequency must be low enough to avoid excessive skin effect. It is noted, however, that this effect is not so serious as might be expected because, when the outer layers pass the Curie point their magnetic coupling is reduced, so that they heat less rapidly than the inner layers. There is therefore a tendency to more uniform temperature generation as the temperature rises.

RESISTANCE BUTT WELDING

When the energy for raising the temperature of the joint is obtained by the resistance to the passage of an electric current across the interface of the joint the process is known as resistance butt welding, upset or just plain butt welding. It would be more accurately described as resistance pressure welding. The process has an intriguing history being probably the first electric-welding method to enjoy serious practical use. It was invented by Professor Elihu Thomson and demonstrated at the American Institute Fair in 1887.

Resistance pressure welding is basically the same as the other processes for pressure butt welding just described. The clamps, however, must apply both force and act as conductors for the low-voltage supply as in flash welding (fig. 75, p. 136). With the work itself, these form part of the single-turn secondary loop of a heavy-duty transformer. Heat is generated initially by the contact resistance, this being dependent on initial force, surface preparation and material composition. Uniform and accurately mating surfaces are desirable to exclude air and give even heating.

The contact resistance decreases as force is raised because asperities and uneven surfaces are deformed on a microscale to increase the effective area of contact. As current passes through the workpiece these points of contact heat up and isolated areas may actually melt, although this is unintentional and undesirable. The contact area continues to increase as the material at the interface becomes plastic so that the contact resistance quickly disappears. By this time, however, there is a band of heated metal with higher resistance than the metal alongside so that the heating process continues (fig. 96). At |this stage the resistivity, temperature coefficient and thermal diffusivity of the metal have a greater effect on resistance and the heating process than pressure. The interfacial extension and particularly the high temperatures reached, which result in diffusion of oxide layers and recrystallization give a high-quality joint.

The hot metal between the clamps may be upset by an increase in the force at the end of the welding cycle or a constant pressure cycle may be adopted, this being frequently employed on small machines.

Resistance pressure welding has been largely displaced for welding cross-sectional areas above 0·25 in. by flash welding which will give higher heat inputs and is therefore more suited for the larger sizes. The process is retained for joining wire and rods and applications where cleanliness and the smooth contour of the upset metal are required. Where high carbon steel is joined it is feasible to give the joint a post-weld heat treatment by passing a second pulse of current through the weld. Welds in these steels are made ductile either by a pulse of current which delays cooling, giving an austempered or annealed weld, or by a delayed pulse which reheats the joint and tempers the structure. Such heat treatments may be done in the welding jaws or a pair of subsidiary contacts.

Diffusion joining

Diffusion welding and diffusion bonding can be considered an extension of the pressure-welding process to higher temperatures and longer times. This enables the welding pressures to be reduced from values sufficient to cause flow to those only high enough to permit intimate contact to be achieved between the parts. The extended times of minutes compared with seconds in pressure welding, at elevated temperatures, require the joint to be made in protective atmosphere or vacuum, the latter also having the advantage of helping to degas the surfaces to be welded. A high temperature, about 1 000 °C for steel, ensures that contact is readily achieved but its main function is to speed up diffusion away from the interface of gas and oxides. With pressure welding conducted at lower temperatures surface films must be ruptured by interfacial extension, but in diffusion welding the films are removed entirely by diffusion.

Diffusion takes place through the volume of a metal by a mechanism involving vacant lattice sites or along grain boundaries, the former mechanism being the more important at elevated temperatures. The process of diffusion is described by the expression

$$D = D_0 e^{-E/RT},$$

where D, the diffusion coefficient, represents the quantity of the solute migrating across a unit cube of solvent in unit time with unit concentration gradient, R is the gas constant and T the temperature in absolute units. D_0 is a constant of the same dimensions as D, and the activation energy E is that necessary to effect motion of atoms from one site to another. The expression illustrates the temperature dependence of the process.

It is found that less energy is required to cause motion of a foreign atom than an atom of the same metal as the one in which the diffusion is taking place. Practical use is made of this in the provision of a thin shim of another metal between two parts to be joined by diffusion bonding. During the process of bonding this layer diffuses into both parts of the workpiece and the final joint may show no evidence of the layer whatever. The film may or may not form a lower melting-point alloy temporarily. Noble metal films are often used and apart from their use in the diffusion process they permit a stable surface to be produced on the workpieces before joining and assist the formation of good contact during the process.

Another variation of the diffusion-bonding process is to braze the joint and then heat-treat the assembly so that the braze metal diffuses into the parent metal and disappears. The advantage of this technique is that the properties of the joint, particularly high-temperature properties, can be greatly improved.

The term 'diffusion welding' has been used to describe the joining of two parts without the addition of a film or shim of another metal. Diffusion bonding covers the techniques where another metal is added, and although this term has been also used for the process involving the heat treatment of a brazed joint it would be more appropriate to use the term 'diffusion brazing'. The group of three processes could be defined collectively as diffusion joining.

Diffusion joining is most satisfactory with the noble metals or those metals which dissolve their own oxides. Diffusion welding has been applied to steels in sizes up to several inches cross-section, while diffusion bonding has been used for joints in reactive and refractory metals. The ability to make joints between dissimilar metals or a metal and a non-metal is a particular attraction of diffusion joining.

FRICTION WELDING

In the friction-welding process the workpieces are brought together under load, one part being revolved against the other so that frictional heat is developed at the interface (fig. 97). When the joint area is sufficiently plastic as a result of the increase in temperature the rotation is halted and the end force increased to forge and consolidate the joint. Friction welding has been used for joining thermo-plastics since 1945, and attempts were made to friction weld metals in the early 1940s by Klopstock. It is to the Russians Chudikov and Vill in 1956, however, that the credit must go for developing a practical method of friction welding metals.

The process takes place in three stages which can be detected on records of the torque applied to the rotating part (fig. 98). Initially the cold parts are subjected to dry friction, but quite rapidly the second phase of the operation begins as a result of local seizures on the faying surface. The number of points of seizure increases and, as each seizure is followed by rupture, the record of torque shows fluctuations. The end of the second stage occurs when maximum torque is reached. In the final phase the torque may drop, the process of seizure and rupture gives way to plastic deformation and conditions reach a steady state. It is in this third phase of the process that most heat is generated—over 85 per cent of the total being a reported value. The yield strength of the metal at the interface is reduced as the temperature rises and falls below the applied shear stress so that plastic metal is extruded from the interface to form a flash or collar round the

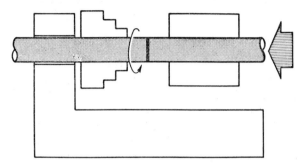

Fig. 97. Principle of friction welding.

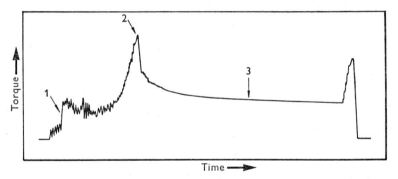

Fig. 98. Torque–time curve for friction welding a carbon steel bar. Arrows indicate the end of the first, second and third stages.

joint. This results in a shortening of the workpieces known as the upset. A peak occurs in the torque curve at the end of the process as a result of speed reduction during braking and forging.

Process characteristics

In the first stage the rubbing surfaces are heated by friction driving off contaminants, asperities are rubbed flat, oxide films are broken and the surfaces are brought to a state when pressure welding can occur. Once the process of seizure and rupture begins the accepted theories of friction no longer apply because the heat is generated by plastic deformation in a layer of finite thickness. Rotational speed variations are reported to have opposite effects in the first two stages and the third stage. Increasing rotational speed with the same end force causes a decrease in the time taken in the first two stages, as might be expected. On the basis of the amount of upset which occurs the third stage is prolonged by increasing speed because the higher rate of heat generation reduces the thickness of the layer of metal which is plastically deformed. The net result of these effects is that for the same upset the total time of welding, stages 1, 2 and 3 combined, passes through a minimum. Summarizing the effects of speed Vill suggests that

$$t_1 + t_2 = A/n, \quad t_3 = Bn + C,$$

where t_1, t_2 and t_3 are the times for stages 1, 2 and 3 respectively; A, B and C are constants and n is the rotational speed in rev/min.

Total heat generation in friction welding is found by experiment to be proportional to the square root of the applied pressure and the power dissipation is independent of speed. The cycle of heat liberation follows the torque curve rising to a maximum and then falling to a steady-state value. All other variables being constant an increase in speed will cause the peak in the torque curve to be passed more quickly and the steady-state condition to be arrived at earlier. There is evidence that once the steady-state condition has been achieved the process can be concluded, and if this criterion rather than fixed attrition of the workpieces is adopted high speeds are found to give more rapid welding than low speeds. There is also a significant reduction in the attrition and the width of the heated zone. For mild steel a peripheral speed of 250–500 ft/min is used and this is independent of the diameter of the stock to be welded over a wide range.

The maximum heat is generated at the periphery of the weld cross-section and zero at the centre where there is no relative motion. An even temperature across the weld is obtained by conduction as the process proceeds. If the process is arrested too soon the heated plastic region is narrow and becomes thinner toward the centre of the workpiece.

Process variables

Speed has little effect on weld quality over a wide range; however, in some steels welds made at high speed may have better impact properties because of the more limited heating. The most noticeable effect of speed is its influence on the shape of the extruded fin. With steel at low speeds the fin is irregular, but at peripheral speeds over about 100 ft/min it becomes uniform. At the highest practicable speeds a minimum amount of metal is extruded into the fin which is thin and convoluted.

Friction pressure is not critical for mild steel being in the range 1–5 tons/in.^{2}. Pressure influences interfacial temperature and torque. An increase in pressure, for example, allows plastic deformation to occur at lower temperatures. The temperature of the plastic interfacial region is therefore reduced while the torque is increased. Usually soft metals, such as aluminium and copper, require low pressures and heat-resisting steels which have good high-temperature properties require high pressures. If high pressures are used on soft metals the joints can become misshapen and twisted. The upset pressures are usually several times the welding pressures.

Each metal or combination of metals appears to have a characteristic torque curve. Metals having high strength at high temperatures give torque–time curves in which the peak between stages 2 and 3 is reduced or missing. This is an indication of reduced plasticity at elevated temperatures compared with mild steel and, since most of the heat is normally generated in stage 3 by plastic deformation, such metals tend to be difficult to weld.

The process is self-regulating as regards temperature in the weld region which rises to just below the melting point at the interface; the precise value of temperature depending on the torque developed and being independent of speed or pressure. Melting is prevented because molten metal cannot transmit stress and acts effectively as a lubricant between the surfaces to be welded. When dissimilar metals are joined the interface temperature is probably just below the lower melting point of the two metals. It is mainly for this reason that the process is excellent for joining dissimilar materials. In the second stage of seizure and rupture with dissimilar metals, a layered structure may develop at the interface. Where the two metals have different ductilities, however, the soft metal may just flow over the surface of the other without the harder metal being greatly affected. This happens with the joint between aluminium and stainless steel. With combinations in which one metal is markedly more plastic at the welding temperature this half of the workpiece may be made a larger diameter so that deformation on this side is reduced.

Control of the process may be by time or by the upset (or shortening) of the workpieces. The latter method tends to be more reliable, especially where low peripheral speeds are used, as the initial stages of the process can proceed at different rates depending on the condition of the joint faces, e.g. roughness or accuracy of presentation. Using upset control the process is not greatly susceptible to surface condition of the faces providing they are not heavily scaled or oxidized.

Applications

Friction welding is used for material from $\frac{3}{16}$ in. diameter up to several inches diameter. The relatively low welding temperatures involved give high-quality joints in many metals including dissimilar combinations. As generally carried out friction welding employs one stationary and one rotating part. It is possible to use a rotating insert between two stationary parts, however, so that two welds are made at once. This enables welds to to be made in long or unwieldy components which cannot be rotated. Contra-rotation is feasible for small diameters but in those sizes other welding processes exist which may be more suitable. At least one and preferably both of the workpieces should be of circular cross-section, either bar or tube. Within these limits the process competes with flash welding, over which it has two advantages—cleanliness and a balanced, steady load on the mains. Power requirements are also greatly reduced compared with flash welding.

FORGE WELDING

For possibly three millennia the only method of welding known was that practised by the blacksmith for joining iron. An early and impressive example of the blacksmith's art is the Iron Pillar of Delhi (pl. 13). Made in A.D. 310 from seven tons of iron blooms, it is 14 in. diameter and stands 24 ft high. In forge welding the parts to be joined are usually heated to over 1 000 °C until they are plastic, and then they are hammered together. Oxide and scale are fluid at the temperatures employed and the hammering is done in such a way as to squeeze this out of the joint and permit metal-to-metal contact. The workpieces may also be rolled, drawn or squeezed together instead of by the application of blows from a hammer. Generally the workpieces are shaped so as to assist the process of squeezing out of oxide and often mechanical interlocking is provided as well.

Wrought iron which contains layers of a readily fused slag is easy to forge weld. Carbon steel requires more precise temperature control and silica sand, fluorspar or borax is sprinkled on the weld faces to make the oxide more fluid. Alloy steels which contain elements giving refractory oxides cannot be welded in this way.

ULTRASONIC WELDING

When two metal workpieces are clamped together between an anvil and a vibrating probe a weld can be produced at the work/work interface. The vibrating probe, called a sonotrode, induces lateral vibration and slip locally between the faying surfaces, disrupting surface films, raising the temperature and forming a type of pressure weld (fig. 99).

Process characteristics

The transfer of energy from the sonotrode to the workpiece it contacts is crucial to the process. There must be no sliding or serious deformation at the tip/work interface while the weld is being made if motion is to be transmitted to the upper of the two workpieces. This condition is affected by such factors as the normal load, amplitude of vibration and the elastic properties of the workpiece. The power required for welding is a function of the normal load, the displacement at the interface and the frequency of vibration. High loads cause deformation of the workpiece, large amplitudes of vibration lead to sliding between tip and work and can cause cracking by fatigue. If energy is to be supplied at a rate necessary to give a weld in a reasonable time, therefore, high frequencies must be used. Frequencies up to 10^5 c/s are used, a common figure being 20 kc/s.

Although the mechanism of ultrasonic welding is not yet completely elucidated it is known to have similarities with friction welding. Initially surface films are dispersed and the contact through asperities at the surface is increased as they break down. Oxide films are ruptured and rolled up by local plastic flow, and dispersion and inter-penetration of the faying surfaces occurs. As in cold welding a controlled surface layer may be helpful

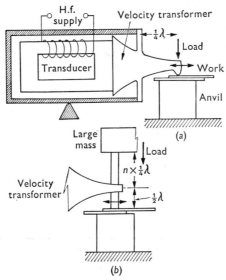

Fig. 99. General arrangement for ultrasonic spot welding: (*a*) direct coupled; (*b*) coupled through a resonant bar.

because chemically clean surfaces are known to produce bonds with negligible inter-penetration. As the temperature rises due to interfacial friction the area of plastic flow extends and welding takes place. With a spot weld the maximum interfacial disturbance occurs in an annulus enclosing a central area where bonding is less complete. The welding process proceeds by the consolidation of this central area within the first annular weld. All this occurs by local slip within an area strained elastically and not by gross sliding. Although the temperature is raised there is no definite evidence of melting, in fact the maximum temperature may be no higher than half the absolute melting point of the metal.

Once a weld has been made and the workpieces are unified, slip must be transferred from the work/work interface to the tip/work interface. Thus, the process depends on the balance between conditions at the tip and those

at the work interface. Temperature measurements give a clue to this behaviour. Russian investigators have reported that a sudden rise in temperature at the work interface occurs at the start of the process, a maximum is
reached and then the temperature falls. If the heat generation is the result
of friction it is to be expected that once a weld is made and slip ceases the
temperature at the work interface will begin to fall. Strength determinations
made on welds in which the process has been arrested at different times
seem to confirm this (fig. 100). The period of no strength at short times

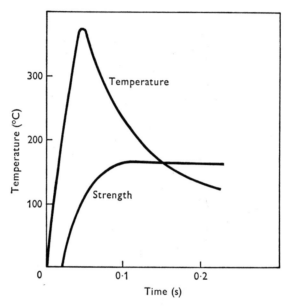

Fig. 100. The development of temperature and strength with time in an
ultrasonic weld (Bolandin and Silin).

corresponds to stage 1 of the friction welding process, the upward curve,
to the stages where local seizure and rupture and growth of the weld occurs.
A similar temperature cycle, but with a maximum at shorter times, is
observed at the tip/work interface. It has been concluded that this corresponds to the cessation of slip at the tip due to an increase in friction
followed by incipient welding of the tip to work. This weak weld must be
ruptured when the workpiece itself is welded so that slip at the tip can occur
once more. The condition of the tips and the upper surface of the work can
therefore have a marked effect on the welding process.

The above explanation is convincing but it must be said that the temperature data in fig. 100 do not agree with measurements by other workers in
the U.S.A. who generally report higher temperatures and no marked drop.
In the Russian work, however, the workpieces themselves were used as the

thermo-couple elements so that the temperatures reported are the average for the whole weld and are significant in indicating a trend in events rather than the actual temperature.

Apparatus for welding

Sonotrode tips are generally made from hardened high-speed steel or Nimonic alloy, materials which have been found to exhibit a low tendency for pressure welding, possibly because of their high strength at elevated temperature. The tips are shaped to present a spherical contour to the work of about 3 in. radius. They may be brazed or welded to the vibrator which supplies the energy for welding.

Ultrasonic vibrators comprise the transducer, which is generally a resonant laminated magnetostrictor and a velocity transformer. The latter is made of a low loss, high strength metal, e.g. titanium, machined to dimensions appropriate to the frequency and material used since $f = \lambda E$, where f is the frequency, λ the wavelength and E the modulus of elasticity. Since the tip must be an antinode the length of the device will be in multiples of $\frac{1}{2}\lambda$ while any supports must be made at the nodal points at $\frac{1}{4}\lambda$. Vibrators must, therefore, operate at one frequency only.

The parts to be welded must be supported on an anvil of sufficient size to prevent the part of the work it contacts from moving in compliance with the vibrations. A device for applying a force between anvil and tip is required and it may be hydraulic, pneumatic or spring-operated according to the size of the unit, springs being used on the smallest equipment. A number of different arrangements of vibrator are possible and two for spot welding are illustrated in fig. 99. For seam welding a disc is machined on the end of the velocity transformer and the whole vibrator is revolved to make the seam. Equipment for making annular welds also employs the tuned reed system with the difference, however, that one or more vibrators are connected tangentially so that the reed vibrates in torsion.

It is difficult to assess how much energy is actually used to make a weld because of unknown losses such as damping within the work or the transmission of energy to the anvil; however, the input power for ultrasonic welders varies from a few watts up to several kilowatts. Clamping forces of a few grammes are used with the former and up to hundreds of pounds with the large units. Power and clamping force are related and depend on the thickness and nature of the material being welded. The power required increases markedly with thickness, for example a tenfold increase in thickness may require a thousandfold increase in power. Less power is required for softer metals. Byron Jones, and others, suggest a relationship as follows between energy required (E), metal thickness (t) and hardness (h):

$$E = Kt^{\frac{3}{2}}h^{\frac{3}{2}},$$

where K is a constant. For each thickness of material there is a particular clamping force at which welds can be made with minimum power, higher or lower forces resulting in increased power demand. With adequate power spot welds are made in less than 1 s.

Applications

Although metal of about $\frac{1}{10}$ in. thickness has been welded by the ultrasonic process it is used chiefly for welding small parts in foil thicknesses. Aluminium welds readily but the process has also been used satisfactorily on a variety of metals and alloys including the refractory metals. Because there is no fusion the method has given good results with dissimilar metals. It is also an ideal method for joining thin to thick workpieces. With the thin metal against the tip the thickness of the part against the anvil does not affect the process. There is very little deformation with the process

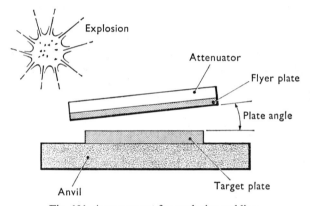

Fig. 101. Arrangement for explosive welding.

except when joining parts such as wires where the contacting area is small. In spot welds indentation need not exceed 5 per cent. The temperature rise in welding is local and the process can be used therefore on temperature-sensitive metals and for delicate components.

EXPLOSIVE WELDING

When two pieces of metal are impacted together a weld can take place at the interface providing certain conditions are met. The velocity of impact required for this type of welding is high but can be achieved by the detonation of an explosive charge. The basic arrangement for explosive welding is indicated in fig. 101.

The lower plate on which welding is to be carried out is laid on a steel anvil and is known as the target plate; the plate which is to be joined to the target plate, often called the projectile or flyer plate, is arranged above the target plate and at a slight angle. To protect the flyer plate from erosion and surface damage and to control the application of the explosive force, a rubber, P.V.C. or similar plate, sometimes known as a buffer or attenuator, is placed between it and the explosive. The angle between the flyer and target plates is normally less than 5° and the flyer plate may be supported away from the target plate by a small distance. Welding is completed in microseconds with a noticeable lack of overall deformation. If a single explosive charge is used it is supported away from the flyer plate by a stand-off distance. When sheet or plastic explosives are used for large area bonding they are placed in contact with the buffer and ignited from the end where the target and flyer plate are closest together. This is called a contact operation.

The technique of explosive welding varies considerably. It has been claimed that an angle between the two plates is essential and that a certain minimum stand-off distance is required. On the other hand, claims of satisfactory welding have also been made where the parts have been brought together parallel or even when directly in contact. The essential feature, however, appears to be that the two surfaces to be joined meet at a slight angle so that a 'bonding front' is established which moves across the interface. Generally the welding operation is conducted in air, although rough vacuum, e.g. 1 torr, has been used. It is claimed that at this pressure lower impacting velocities are required as there is no air cushion to overcome.

It is not necessary to prepare the surfaces of metal for welding by explosive forces. However, deeply pitted or scaly surfaces should be avoided, as also should be heavily sand-blasted or roughened surfaces. Satisfactory welds can be made between copper and steel and a variety of metals such as gold, silver, nickel and titanium.

Explosive welding is primarily a method of joining large overlapping areas in which the flyer is sheet material. The plate sizes employed have varied from a few square inches up to several yards square. Because at the edge of a plate the pressure is relieved, there is a tendency for the impact velocity in these regions to be reduced. It is possible therefore when welding small plates at near minimum velocities to find that welding is incomplete around the edges.

Nature of the bond

There is considerable plastic deformation in the immediate region of the faying surface such that the hardness of the deformed interface is usually greater than for unwelded sheets. The interface between the welded workpieces invariably takes a wavy form, the amplitude of the waves being from 0·005 to $\frac{3}{16}$ in. with wavelengths from 0·010 to $\frac{1}{4}$ in. depending on the welding conditions. This formation of waves appears to be a requirement for a satisfactory weld. A phenomenon known as surface jetting is also observed which takes the form of a small jet of metal ejected from between the plates as they fold together.

The mechanism by which the ripples and jet are formed is that under the high velocity and therefore high pressure and instantaneous temperatures of explosive impact the metal in the immediate region of the bonding front is sufficiently plastic to behave as a fluid. Under these conditions the flyer plate resembles a jet which is traversed across the target plate. A hump is raised in the plastic region in the target ahead of the jet which periodically passes over the hump and forms another further on. This explains the formation of a rippled weld interface. When a subsonic jet is directed against a near-solid surface the jet divides, the relative proportions of the forward and backward flow depending on the angle at which the jet meets

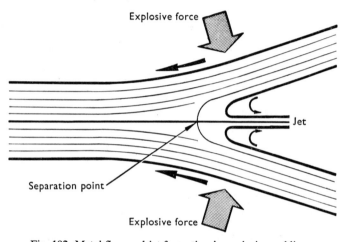

Fig. 102. Metal flow and jet formation in explosive welding.

the solid. The division of the jet in this way accounts for the observations in explosive welding of the surface jetting and the crests as shown in pl. 14 which are often observed on the ripples at the weld interface. The division of the jet results in a separation point as indicated in fig. 102 so that the extreme surface of the contacting plates is stripped off, exposing metal in an ideal manner for bonding. Metal and surface films removed in this way are present in the crests to the waves. These conditions are ideal for welding; however, bonding may still take place when surface jetting is not obvious because of the surface extension in rippling.

Velocity of impact

Impact velocity depends on the ratio of the weight of the charge to that of the flyer and also on contact angle. There is evidence that there is a minimum velocity below which welding does not take place, for example satisfactory welds cannot be made in copper at velocities of less than 400 ft/s or in aluminium at velocities less than 750 ft/s. This minimum velocity could be determined by the point at which the projectile material becomes sufficiently plastic on impact to form a divided jet. In spite of the suggestion that the process is akin to cold-pressure welding, it is hard to believe that, locally at any rate, quite high temperatures are not achieved. Joints between copper and aluminium, for example, can often show distinct indications of alloying at the interface giving a brittle joint. Such excessive heating may well be a result of using unnecessarily high impacting velocities. The maximum velocity which it is practical to use is fixed by the velocity of sound in the target material since at supersonic velocities the wave in the target cannot propagate ahead of the bonding front.

Angle of contact

The recommended angle between projectile and target plates prior to welding is between 2° and 4°. Although welds can be made with angles several times this the plates should be as near parallel as possible to give nearly equal jet and base velocities and thus prevent sheet separation and erosion.

Applications

Explosive welding is a specialized process which can only be applied to variations of the lap joint. It opens the possibility, however, of making joints of a type which cannot be produced easily any other way. Cladding down to foil gauges is perhaps the most attractive application and variations of this technique would indicate the fabrication of heat exchangers by welding sheets to plates in which channels had been machined. Sleeved joints in tube and tube to tube-plate joints are also feasible when the charge is exploded within the tube. The process could also be employed for making welds in places inaccessible to conventional processes or for site welding where power and skill for fusion welding are difficult to obtain.

Bibliography

Agers, B. M. and Singer, A. R. E. (1964). The mechanism of small tool pressure welding. *Br. Weld. J.* **11**, no. 7, 313–19.

Bahrani, A. S. and Crossland, B. (1964–5). Explosive welding and cladding. *Inst. Mech. Eng. Proc.* **179**, pt. 1, no. 7, 264–81.

Balandin, G. F. and Silin, L. L. (1960). The role of friction in ultrasonic welding. *Russian Metallurgy and Fuels.* D.S.I.R. Nat. Lending Lib. RO 220-7815, no. 6.

Baranov, I. B. (1959). *Cold Welding of Ductile Metals.* Translation, D.S.I.R. Nat. Lending Lib. Boston Spa.

Brown, D. C. and Wilson, J. J. (1955). Oxy-acetylene pressure welding of aircraft undercarriage components. *Br. Weld. J.* **2**, no. 4, 160–71.

Conti, R. J. (1966). Thermo-compression joining techniques for electronic devices and interconnects. *Metals Eng. Quart.* **6**, no. 1, 29–35.

Curran, G. R. and Holtzman, A. H. (1963). Flow configurations in colliding plates: explosive bonding. *J. appl. Phys.* **34**, no. 4, pt. 1, 928–39.

Forging by the blacksmith (1939). *A.S.M. Metals Handbook.*

Gerken, J. M. and Owczarski, W. A. (1965). A review of diffusion welding. *Weld. Res. Counc. Bull.* no. 109.

Harris, S. G. (1961). Pressure butt welding of steel pipe using induction heating. *Weld. J.* **40**, no. 2, 57S–65S.

Hazlett, T. H. and Gupta, K. K. (1963). Friction welding of high strength aluminium alloys. *Weld. J.* **42**, no. 11, 490S–4S.

Hollander, M. B., Chengard, C. J. and Wyman, J. C. (1963). Friction welding parameter analysis. *Weld. J.* **42**, no. 11, 495S–501S.

Klopstock, H. and Neelands, A. R. (1941). Brit. Pat. 572,789.

Jones, J. B., Maropis, M., Thomas, J. G. and Bancroft, D. (1961). Phenomenologica considerations in ultrasonic welding. *Weld. J.* **40**, no. 7, 289S–305S.

Milner, D. R. and Rowe, G. W. (1962). Fundamentals of solid-phase welding. *Metallurgical Rev.* **7**, no. 28, 433–80.

Pocalyko, A. and Williams, C. P. (1964). Clad plate products by explosion bonding. *Weld. J.* **43**, no. 10, 854–61.

Vill, V. I. (1962). *Friction Welding of Metals.* Translation, American Weld. Soc.

Voinov, V. P. and Kupershlyak-Yuzefovich, G. M. (1964). Relative rotational speed in friction welding. *Welding Production*, no. 3, 15–20. (Translation of Svar. Proiz by B.W.R.A.)

INDEX